Student Solutions Manual
for

Modern Physics
Fourth Edition

PAUL A. TIPLER
RALPH A. LLEWELLYN

Prepared by
MARK J. LLEWELLYN
School of Electrical Engineering and Computer Science
University of Central Florida

W. H. Freeman and Company
New York

ISBN: 0-7167-9844-1

Table of Contents

Preface

This book is a Student Solutions Manual for the end-of-chapter problems in *Modern Physics, 4e* by Paul A. Tipler and Ralph A. Llewellyn. This book contains solutions to every fourth end-of-chapter problem in the text.

Figure numbers, equations, and table numbers refer to those in the text. Figures in this solutions manual are not numbered and correspond only to the problem in which they appear. Notation and units parallel those in the text.

Please visit the W. H. Freeman Physics Web site at www.whfreeman.com/modphysics4e where you will find Chapter 14 of *Modern Physics, 4e* and the accompanying MORE sections for each chapter.

Every effort has been made to ensure that the solutions in this manual are accurate and free from errors. If you have found an error or a better solution to any of these problems, please feel free to contact me at the address below with a specific citation. I appreciate any correspondence from users of this manual who have ideas and suggestions for improving it.

Sincerely,

Mark J. Llewellyn
School of Electrical Engineering
University of Central Florida
Orlando, Florida 32186-2362
email: mark@cs.ucf.edu

Chapter 1 − Relativity I

1-1. Once airborne, the plane's motion is relative to still air. In 10 min the air mass has moved $18\,m/s \times 60\,s/min \times 10\,min = 10.8\,km$ toward the east. The north and up coordinates relative to the ground (and perpendicular to the wind direction) are unaffected. The 25 km point has moved 10.8 km east and is, after 10 min, at $25 - 10.8 = 14.2\,km$ west of where the plane left the ground (0, 0, 0) after 10 min the plane is at (14.2 km, 16 km, 0.5 km).

1-5. (a) In this case, the situation is analogous to Example 1-3 with $L = 3 \times 10^8\,m$, $v = 3 \times 10^4\,m/s$, and $c = 3 \times 10^8\,m/s$. If the flash occurs at $t = 0$, the interior is dark until $t = 2$ s at which time a bright circle of light reflected from the circumference of the great circle plane perpendicular to the direction of motion reaches the center, the circle splits in two, one moving toward the front and the other toward the rear, their radii decreasing to just a point when they reach the axis 10^{-8} s after arrival of the first reflected light ring. Then the interior is again dark.

  (b) In the frame of the seated observer, the spherical wave expands outward at c in all directions.

  (g) Yes. The charge is an intrinsic property of the electron, a fundamental constant.

1-9. The wave from the front travels 500 m at speed $c + (150/3.6)\,m/s$ and the wave from the rear travels at $c - (150/3.6)\,m/s$. As seen in Figure 1-15, the travel time is longer for the wave from the rear.

(Problem 1-9 continued)

$$\Delta t = t_r - t_f = \frac{500\,m}{3.00\times10^8\,m/s - (150/3.6)\,m/s} - \frac{500\,m}{3.00\times10^8\,m/s + (150/3.6)\,m/s}$$

$$= 500\left[\frac{3\times10^8 + (150/3.6) - 3\times10^8 + (150/3.6)}{(3\times10^8)^2 - 2(150/3.6)(3\times10^8) - (150/3.6)^2}\right]$$

$$\approx 500\frac{2(150/3.6)}{(3\times10^8)^2} \approx 4.63\times10^{-13}\,s$$

1-13. (a) $\gamma = 1/(1-v^2/c^2)^{1/2} = 1/[1-(0.85c)^2/c^2]^{1/2} = 1.898$

$x' = \gamma(x-vt) = 1.898[75\,m - (0.85c)(2.0\times10^{-5}s)] = -9.537\times10^3\,m$

$y' = y = 18\,m$

$z' = z = 4.0\,m$

$t' = \gamma(t-vx/c^2) = 1.898[2.0\times10^{-5}s - (0.85c)(75\,m)/c^2] = 3.756\times10^{-5}\,m$

(b) $x = \gamma(x'+vt') = 1.898[-9.537\times10^3\,m + (0.85c)(3.756\times10^{-5}s)] = 75.8$

(difference is due to rounding of γ, x', and t'.

$y = y' = 18\,m$

$z = z' = 4.0\,m$

$t = \gamma(t' + vx'/c^2) = 1.898[3.756\times10^{-5}s + (0.85c)(-9.537\times10^3\,m)/c^2]$

$= 2.0\times10^{-5}\,s$

1-17. (a) As seen from the diagram, when the observer in the rocket (S') system sees 1 c·s tick by on the rocket's clock, only 0.6 c·s have ticked by on the laboratory clock.

(b) When 10 seconds have passed on the rocket's clock, only 6 seconds have passed on the laboratory clock.

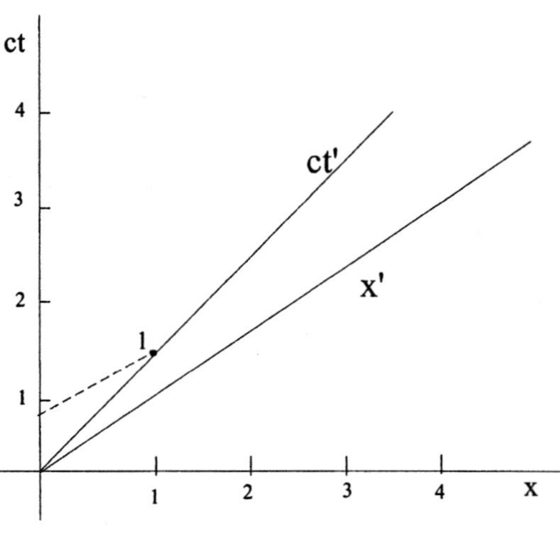

2

1-21. $\quad \Delta t = \gamma \, \Delta t' \quad$ (Equation 1-28)

$$\frac{\Delta t - \Delta t'}{\Delta t'} = \frac{\gamma \, \Delta t' - \Delta t'}{\Delta t'} = \gamma - 1 \approx \frac{1}{2}\frac{v^2}{c^2} = \gamma - 1 \approx \frac{1}{2}\frac{v^2}{c^2}$$

$$v^2 = 2c^2\frac{\Delta t - \Delta t'}{\Delta t'} \qquad v = c\left(2 \times \frac{\Delta t - \Delta t'}{\Delta t'}\right)^{\frac{1}{2}} = c(2 \times 0.01)^{\frac{1}{2}} = 0.14\,c$$

1-25. From Equation 1-30, $L = L_p/\gamma = L_p\sqrt{1 - v^2/c^2}$

where $L = 85\text{m}$ and $L_p = 100\text{m}$

$$\sqrt{1 - v^2/c^2} = L/L_p = 85/100$$

$$\textit{Squaring} \quad 1 - v^2/c^2 = (85/100)^2$$

$$\therefore \quad v^2 = \left[1 - (85/100)^2\right]c^2 = 0.2775\,c^2 \ \textit{and} \ v = 0.527c = 1.58 \times 10^8\,m/s$$

1-29. (a) In S': $V' = a' \times b' \times c' = (2m)(2m)(4m) = 16\,m^3$

In S: Both a' and c' have components in the x' direction.

$$a_x' = a'\sin 25° = (2m)\sin 25° = 0.84\,m \ \text{ and } \ c_x' = c'\cos 25° = (4m)\cos 25° = 3.63\,m$$

$$a_x = a_x'\sqrt{1 - \beta^2} = 0.84\sqrt{1 - (0.65)^2} = 0.64\,m$$

$$c_x = c_x'\sqrt{1 - \beta^2} = 3.634\sqrt{1 - (0.65)^2} = 2.76\,m$$

$$a_y = a_y' = a'\cos 25° = 2\cos 25° = 1.81\,m \ \text{ and } \ c_y = c_y' = c'\sin 25° = 4\sin 25° = 1.69\,m$$

$$a = \sqrt{a_x^2 + a_y^2} = \sqrt{(0.64)^2 + (1.81)^2} = 1.92\,m$$

$$c = \sqrt{c_x^2 + c_y^2} = \sqrt{(2.76)^2 + (1.69)^2} = 3.24\,m$$

b' (in z direction) is unchanged, so $b = b' = 2m$

θ (between c and xy-plane) $= \tan^{-1}(1.69/2.76) = 31.5°$

(Problem 1-29 continued)

ϕ (between a and yz-plane) $= \tan^{-1}(0.64/1.81) = 19.5°$

$V = $ (area of ay face) $\cdot b$ (see part [b])

$$V = (c \times a \sin 78°) \times b = (3.24\,m)(1.92\,m \sin 78°)(2\,m) = 12.2\,m^3$$

(b)

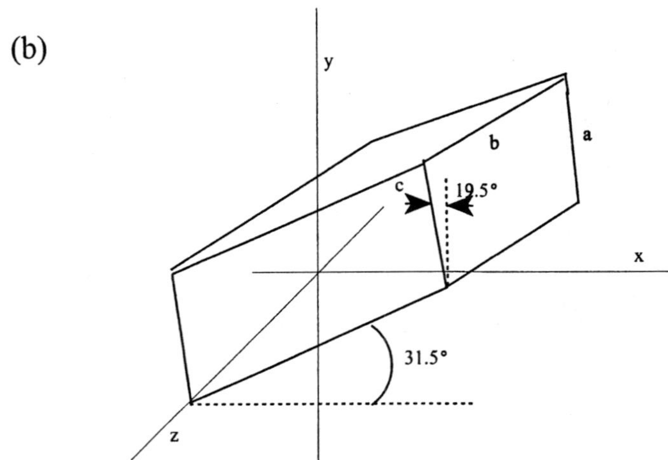

1-33. $f = \sqrt{\dfrac{1-\beta}{1+\beta}}\, f_0 \quad \rightarrow \quad \lambda = \sqrt{\dfrac{1-\beta}{1+\beta}}\, \lambda_0 = \sqrt{\dfrac{1-\beta}{1+\beta}}\,(656.3\ nm)$

For $\beta = 10^{-3}$: $\lambda = (656.3\ nm)\sqrt{\dfrac{1+10^{-3}}{1-10^{-3}}} = 657.0\ nm$

For $\beta = 10^{-2}$: $\lambda = (656.3\ nm)\sqrt{\dfrac{1+10^{-2}}{1-10^{-2}}} = 662.9\ nm$

For $\beta = 10^{-1}$: $\lambda = (656.3\ nm)\sqrt{\dfrac{1+10^{-1}}{1-10^{-1}}} = 725.6\ nm$

1-37. $\cos\theta = \dfrac{\cos\theta' + \beta}{1 + \beta\cos\theta'}$ (Equation 1-43)

where θ' = half-angle of the beam in $S' = 30°$

For $\beta = 0.65$, $\cos\theta = \dfrac{\cos 30° + 0.65}{1 + (0.65)\cos 30°} = 0.97$ or $\theta = 1.41°$

The train is A from you when the headlight disappears, where

$$A = \frac{0.75\,m}{\tan 14.1°} = 3.0\,m$$

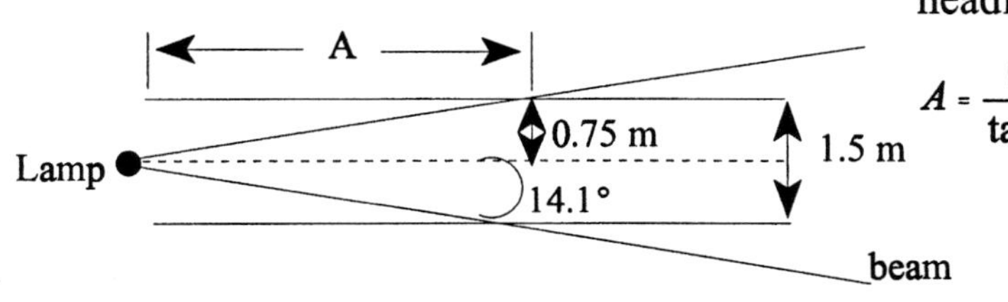

1-41. Orbit circumference $\approx 4.0 \times 10^7\,m$.

Satellite speed $v = 4.0 \times 10^7\,m/(90\,min \times 60\,s/min) = 7.41 \times 10^3\,m/s$

$\Delta t - \Delta t_0 = t_{diff}$

$\Delta t - \Delta t/\gamma = t_{diff} = \Delta t(1 - 1/\gamma) \approx \Delta t\left(\dfrac{1}{2}\beta^2\right)$ (Problem 1-20)

$t_{diff} = (3.16 \times 10^7\,m/s)(1/2)(7.41 \times 10^3/3.0 \times 10^8)^2$

$= 0.0096\,s = 9.6\,ms$

1-45. (a)

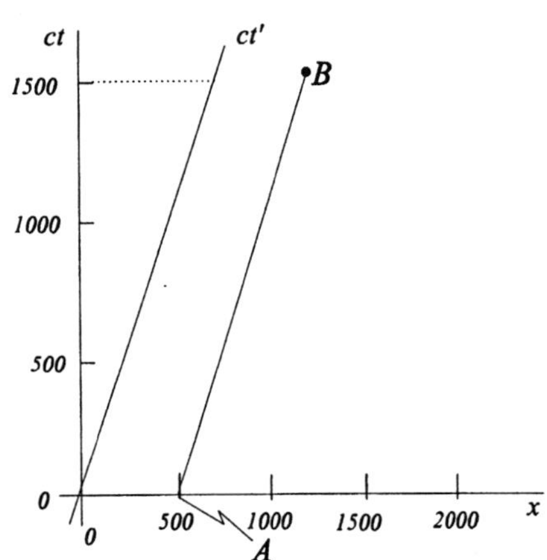

(Problem 1-45 continued)

(b) Slope of ct' axis $= 2.08 = 1/\beta$, so $\beta = 0.48$ and $v = 1.44 \times 10^8 \ m/s$

(c) $ct' = \gamma ct$ and $\gamma = 1/\sqrt{1-\beta^2}$ so $ct'\sqrt{1-\beta^2} = ct$

For $ct' = 1000 \, m$ and $\beta = 0.48$ $ct = 877 \, m$

(d) $\Delta t = \gamma \Delta t' = 1.14 \Delta t' \rightarrow \Delta t' = 5\,\mu s/1.14 = 4.39 \,\mu s$

1-49. $t_2' - t_1' = \gamma\left(t_2 - t_1\right) - \dfrac{\gamma v}{c^2}\left(x_b - x_a\right)$ (Equation 1-22)

(a) $t_2' - t_1' = 0 \rightarrow t_2 - t_1 = \left(v/c^2\right)\left(x_b - x_a\right) \rightarrow (0.5 - 1.0)y = \left(v/c^2\right)(2.0 - 1.0)\,c\cdot y$

Thus, $-0.5 = (v/c) \rightarrow v = 0.5c$ in the $-x$ direction.

(b) $t' = \gamma\left(t - vx/c^2\right)$

Using the first event to calculate t' (because t' is the same for both events),

$t = \left(1/\sqrt{1-(0.5)^2}\right)\left[1y - (0.5c)(1c\cdot y)/c^2\right] = 1.155(0.5) = 0.58 \, y$

(c) $(\Delta s)^2 = (\Delta x)^2 - (c\Delta t)^2 = (1c\cdot y)^2 - (0.5c\cdot y)^2 = 0.75(c\cdot y)^2 \rightarrow \Delta s = 0.866 c \cdot y$

(d) The interval is spacelike.

(e) $L_p = \Delta s = 0.866 c \cdot y$

1-53. This is easier to do in the xy and $x'\,y'$ planes. Let the center of the meterstick, which is parallel to the x axis and moves upward with speed v_y in S, at $x = y = x' = y' = 0$ at $t = t' = 0$. The right hand end of the stick, e.g., will not be at $t' = 0$ in S' because the clocks in S' are not synchronized with those in S. In S' the components of the sticks velocity are:

$u_y' = \dfrac{u_y}{\gamma\left(1 - vu_x/c^2\right)} = \dfrac{v_y}{\gamma}$ because $u_y = v_y$ and $u_x = 0$

$u_x' = \dfrac{u_x - v}{1 - vu_x/c^2} = -v$ because $u_x = 0$

(Problem 1-53 continued)

When the center of the stick is located as noted above, the right end in S' will be at:

$x' = \gamma(x - vt) = 0.5\gamma$ because $t = 0$. The S' clock there will read:

$t' = \gamma(t - vx/c^2) = -0.5\gamma v/c^2$ because $t = 0$. Therefore, when $t' = 0$ at the center, the right end is at $x'\,y'$ given by:

$$x' = 0.5\,\gamma \qquad y' = u_y' t' = \frac{v_y}{y}\left(\frac{0.5\,\gamma v}{c^2}\right)$$

$$\text{and } \theta' = \tan^{-1}\frac{y'}{x'} = \frac{v_y}{\gamma}\left(\frac{0.5\,\gamma v}{c^2}\right)\Big/ 0.5\,\gamma = \left(v_y v/c^2\right)\sqrt{1 - \beta^2}$$

$$\text{For } \beta = 0.65 \qquad \theta' = \left(0.494\, v_y/c\right)$$

1-57. $v = 0.6c$ $\qquad \gamma = \dfrac{1}{\sqrt{1 - (0.6)^2}} = 1.25$

(a) The clock in S reads $\gamma \times 60\,\text{min} = 75\,\text{min}$ when the S' clock reads 60 min and the first signal from S' is sent. At that time, the S' observer is at $v \times 75\,\text{min} = 0.6c \times 75\,\text{min} = 45\,c\cdot\text{min}$. The signal travels for 45 min to reach the S observer and arrives at 75 min + 45 min = 120 min on the S clock.

(b) The observer in S sends his first signal at 60 min and its subsequent wavefront is found at $x = c(t - 60\,\text{min})$. The S' observer is at $x = vt = 0.6ct$ and receives the wavefront when these x positions coincide, i.e., when

$$\begin{aligned}
c(t - 60\,\text{min}) &= 0.6ct \\
0.4ct &= 60\,c\cdot\text{min} \\
t &= (60\,c\cdot\text{min})/0.4c = 150\,\text{min} \\
x &= 0.6c(150\,\text{min}) = 90\,c\cdot\text{min}
\end{aligned}$$

The confirmation signal sent by the S' observer is sent at that time and place, taking 90 min to reach the observer in S. It arrives at 150 min + 90 min = 240 min.

(Problem 1-57 continued)

(c) Observer in S:

Sends first signal	60 min
Receives first signal	120 min
Receives confirmation	240 min

The S' observer makes identical observations.

1-61. (a) Apparent time $A \to B = T/2 - t_A + t_B$ and apparent time $B \to A = T/2 + t_A - t_B$

where t_A = light travel time from point A to Earth and t_B = light travel time from point B to Earth.

$$A \to B = \frac{T}{2} - \frac{L}{c+v} + \frac{L}{c-v} = \frac{T}{2} + \frac{2vL}{c^2 - v^2}$$

$$B \to A = \frac{T}{2} - \frac{L}{c+v} - \frac{L}{c-v} = \frac{T}{2} - \frac{2vL}{c^2 - v^2}$$

(b) Star will appear at A and B simultaneously when $t_B = T/2 + t_A$

or when the period is:

$$T = 2\left[t_B - t_A\right] = 2\left[\frac{L}{c-v} - \frac{L}{c+v}\right] = \frac{4vL}{c^2 - v^2}$$

2-1. $u_{yB}^2 = u_0^2\left(1 - v^2/c^2\right)$ $u_{xB}^2 = v^2$

$$\sqrt{1 - \left(u_{xB}^2 + u_{yB}^2\right)\!/c^2} = \sqrt{1 - v^2/c^2 - \left(u_0^2/c^2\right)\!\left(1 - v^2/c^2\right)}$$

$$= \sqrt{\left(1 - v^2/c^2\right)\!\left(1 - u_0^2/c^2\right)}$$

$$= \left(1 - v^2/c^2\right)^{1/2}\left(1 - u_0^2/c^2\right)^{1/2}$$

$$p_{yB} = \frac{m\,u_{yB}}{\sqrt{1 - \left(u_{xB}^2 + u_{yB}^2\right)\!/c^2}} = \frac{-\,m u_0\sqrt{1 - v^2/c^2}}{\sqrt{1 - v^2/c^2}\,\sqrt{1 - u_0^2/c^2}}$$

$$= -\,m u_0 \Big/ \sqrt{1 - u_0^2/c^2} = -p_{yA}$$

2-5. $\Delta E = \Delta m c^2$ $\therefore$ $\Delta m = \Delta E/c^2 = \dfrac{10\,J}{(3.0\times 10^8\,m/s)^2} = 1.1\times 10^{-16}\,kg$

Because work is done *on* the system, the mass *increases* by this amount.

2-9. $E = \gamma m c^2$ (Equation 2-10)

(a) $200\,GeV \times 197 = 3.94\times 10^4\,GeV = \gamma\,(0.938\,GeV)$ where $mc^2(\text{proton}) = 0.938\,GeV$

$$\gamma = \frac{1}{\sqrt{1 - v^2/c^2}} = \frac{3.94\times 10^4\,GeV}{0.938\,GeV} = 4.20\times 10^4 \quad \text{Because } E \gg mc^2,$$

$$\frac{v}{c} \approx 1 - \frac{1}{2\gamma^2} \quad \text{(Equation 2-40)}$$

$$\frac{v}{c} \approx 1 - \frac{1}{2\left(4.20\times 10^4\right)^2} = 1 - 0.0000000002834 \quad \text{Thus, } v \approx 0.999999999717\,c$$

(b) $E \approx pc$ for $E \gg mc^2$ (Equation 2-36)

$p = E/c = 3.94\times 10^4\,GeV/c$

(c)   Assuming one *Au* nucleus (system S') to be moving in the $+x$ direction of the lab (system S), then u for the second *Au* nucleus is in the $-x$ direction. The second *Au's* energy measured in the S' system is :

(Problem 2-9 continued)

$$E = \gamma\left(E + vp_x\right) = \left(4.20 \times 10^4\right)\left(3.94 \times 10^4\, GeV + v\, 3.94 \times 10^4\, GeV/c\right)$$

$$= \left(4.20 \times 10^4\right)\left(3.94 \times 10^4\, GeV\right)\left(1 + v/c\right)$$

$$= \left(4.20 \times 10^4\right)\left(3.94 \times 10^4\, GeV\right)\left(2\right)$$

$$= 3.31 \times 10^9\, GeV$$

$$p_x = \gamma\left(p_x - vE/c^2\right) = \left(4.20 \times 10^4\right)\left(- 3.94 \times 10^4\, GeV - v\, 3.94 \times 10^4\, GeV/c^2\right)$$

$$= -\left(4.20 \times 10^4\right)\left(3.94 \times 10^4\, GeV\right)\left(2\right)$$

$$= - 3.31 \times 10^9\, GeV/c$$

2-13. (a) $\gamma = E/mc^2 = 2 = 1/\sqrt{1 - u^2/c^2} \;\rightarrow\; 1 - u^2/c^2 = 1/4$

$$\rightarrow\; u^2/c^2 = 1 - 1/4 = 0.75 \;\rightarrow\; u = 0.866\,c$$

(b) $p = \dfrac{1}{c}\left[E^2 - \left(mc^2\right)^2\right]^{1/2}$ (From Equation 2-32)

$$p = \frac{1}{c}\left[\left(2mc^2\right)^2 - \left(mc^2\right)^2\right]^{1/2} = \frac{1}{c}\left[4\left(mc^2\right)^2 - \left(mc^2\right)^2\right]^{1/2} = \sqrt{3}\; mc$$

2-17. $^3H \;\rightarrow\; ^2H + n$

Energy to remove the n

$$= 22.014102\,u\left(^2H\right) + 1.008665\,u(n) - 3.016049\,u\left(^3H\right)$$

$$= 0.006718\,u \times 931.5\, MeV/u = 6.26\, MeV$$

2-21. Conservation of energy requires that $E_i^2 = E_f^2$, or

$$\left(p_i c\right)^2 + \left(2 m_p c^2\right)^2 = \left(p_f c\right)^2 + \left(2 m_p c^2 + m_\pi c^2\right)^2$$ and conservation of momentum requires that

$p_i = p_f$, so

(Problem 2-21 continued)

$$4\left(m_p c^2\right)^2 = 4\left(m_p c^2\right)^2 + 2 m_p c^2 \times 2 m_\pi c^2 + \left(m_\pi c^2\right)^2$$

$$0 = 2 m_p c^2 \times 2 m_\pi c^2 + \left(m_\pi c^2\right)^2$$

$$0 = m_\pi c^2 \left(2 + \frac{m_\pi c^2}{2 m_p c^2}\right) = m_\pi c^2 \left(2 + \frac{m_\pi}{2 m_p}\right)$$

Thus, $m_\pi c^2 \left(2 + m_\pi / 2 m_p\right)$ is the minimum or threshold energy E_t that a beam proton must have to produce a π^0.

$$E_t = m_\pi c^2 \left(2 + \frac{m_\pi c^2}{2 m_p c^2}\right) = 135\,MeV\left(2 + \frac{135}{2(938)}\right) = 280\,MeV$$

2-25.   Positronium at rest: $\left(2 m c^2\right)^2 = E_i^2 + \left(p_i c\right)^2$

Because $p_i = 0$, $E_i = 2 m c^2 = 2(0.511\,MeV) = 1.022\,MeV$

After photon creation; $\left(2 m c^2\right)^2 = E_f^2 + \left(p_f c\right)^2$

Because $p_f = 0$ and energy is conserved, $\left(2 m c^2\right)^2 = E_f^2 = (1.022\,MeV)^2$ or $2 m c^2 = 1.022\,MeV$ for the photons.

2-29. $E^2 = (pc)^2 + \left(m c^2\right)^2$ (Equation 2-31)

$$(1746\,MeV)^2 = (500\,MeV)^2 + \left(m c^2\right)^2$$

$$m c^2 = \left[(1746\,MeV)^2 - (500\,MeV)^2\right]^{1/2} = 1673\,MeV \rightarrow m = 1673\,MeV/c^2$$

$$E = \gamma m c^2 \rightarrow \gamma = 1/\sqrt{1 - u^2/c^2} = E/m c^2$$

$$u/c = \left[1 - \left(m c^2/E\right)^2\right]^{1/2} = \left[1 - (1673\,MeV/1746\,MeV)^2\right]^{1/2} = 0.286 \rightarrow u = 0.286 c$$

2-33. Because the clock furthest from Earth (where Earth's gravity is less) runs the faster, answer (c) is correct.

2-37. The speed v of the satellite is:

$$v = 2\pi R/T = 2\pi(6.37\times 10^6\,m)/(90\min \times 60\,s/\min) = 7.42\times 10^3\,m/s$$

Special relativistic effect:

After one year the clock in orbit has recorded time $\Delta t = \Delta t/\gamma$, and the clocks differ by:

$$\Delta t - \Delta t = \Delta t - \Delta t/\gamma = \Delta t(1-1/\gamma) \approx \Delta t(v^2/2c^2), \text{ because } v << c. \text{ Thus,}$$

$$\Delta t - \Delta t = (3.16\times 10^7\,s)(7.412\times 10^3)^2/(2)(3.0\times 10^8\,m)^2 = 0.00965\,s = 9.65\,ms$$

Due to special relativity time dilation the orbiting clock is behind the Earth clock by 9.65 *ms*.

General relativistic effect:

$$\frac{\Delta f}{f_0} = \frac{gh}{c^2} = \frac{(9.8\,m/s^2)(3.0\times 10^5\,m)}{(3.0\times 10^8\,m/s)^2} = 3.27\times 10^{-11}\,s/s$$

In one year the orbiting clock gains

$$(3.27\times 10^{-11}\,s/s)(3.16\times 10^7\,s/y) = 1.03\,ms$$

The net difference due to both effects is a slowing of the orbiting clock by $9.65 - 1.03 = $ 8.62 *ms*.

2-41. (a) The momentum p of the ejected fuel is:

$$p = \gamma mu = mu/\sqrt{1-u^2/c^2} = 10^3\,kg(c/2)/\sqrt{1-0.5^2} = 1.73\times 10^{11}\,kg\cdot m/s$$

Conservation of momentum requires that this also be the momentum p_s of the spaceship:

$$p_s = m_s u_s/\sqrt{1-u_s^2/c^2} = 1.73\times 10^{11}\,kg\cdot m/s$$

Or, $m_s u_s/\sqrt{1-u_s^2/c^2} = (1.73\times 10^{11}\,kg\cdot m/s)^2$

$$m_s^2 c_s^2 = (1-u_s^2/c^2)(1.73\times 10^{11}\,kg\cdot m/s)^2 = (1.73\times 10^{11}\,kg\cdot m/s^2) - (3.33\times 10^5\,kg^2)u_s^2$$

$$(10^6\,kg)^2 u_s^2 + (3.33\times 10^5\,kg^2)u_s^2 = (1.73\times 10^{11}\,kg\cdot m/s)^2$$

Or, $u_s = (1.73\times 10^{11}\,kg\cdot m/s)/10^6\,kg = 1.73\times 10^5\,m/s = 5.77\times 10^{-4}c$

(Problem 2-41 continued)

(b)   In classical mechanics, the momentum of the ejected fuel is :

$mu = mc/2 = 10^3 c/2$, which must equal the magnitude of the spaceship's

momentum $m_s u_s$, so

$$u_s = 10^3 (c/2)/m_s = \frac{10^3\, kg\,(3.0 \times 10^8\, m s/)}{2(10^6\, kg)} = 5.0 \times 10^{-4} c = 1.5 \times 10^5\ m/s$$

(c)   The initial energy E_i before the fuel was ejected is $E_i = m_s c^2$ in the ship's rest

system. Following fuel ejection, the final energy E_f is:

$$E_f = \text{energy of fuel} + \text{energy of ship} = mc^2/\sqrt{1-u^2/c^2} + (m_s - m)c^2/\sqrt{1-u_s^2/c^2}$$

where $u = 0.5c$ and $u_s << c$, so

$$E_f = 1.155\, mc^2 + (m_s - m)c^2 = (1.155 - 1)mc^2 + m_s c^2$$

The change in energy ΔE is

$$\Delta E = E_f - E_i = \left[(0.155)(10^3\, kg)c^2 + 10^6\, kg c^2\right] - \left[10^6\, kg c^2\right]$$

$\Delta E = 155\, kg c^2$ or $155\, kg = \Delta E/c^2$ of mass has been converted to energy.

2-45. $$p_y = \gamma\, mu_y = \left[\frac{\gamma\left(1 - vu_x/c^2\right)}{\sqrt{1-u^2/c^2}}\right] \times m \times \left[\frac{u_y}{\gamma\left(1 - u_x v/c^2\right)}\right]$$

Canceling γ and $\left(1 - vu_x/c^2\right)$, gives: $p_y = \dfrac{m u_y}{\sqrt{1-u^2/c^2}} = P_y$

In an exactly equivalent way, $p_z = P_z$.

2-49.   (a)   If v mass is 0:

$$E_\mu^2 = (p_\mu c)^2 + (m_\mu c)^2 \quad \text{and} \quad E_v^2 = (p_v c^2)^2 + 0$$

$$E_{k\mu} + E_v = 139.56755\, MeV - 105.65839\, MeV$$

$$m_\mu c^2 (\gamma - 1) + E_v = 33.90916\, MeV$$

$$p_\mu c = \left(E_\mu^2 - (m_\mu c^2)^2\right)^{1/2} = 33.90916 - m_\mu c^2 (\gamma - 1)$$

Squaring, we have

(Problem 2-49 continued)

$$\left(m_\mu c^2\right)^2\left(\gamma^2-1\right) = (33.90916)^2 - 2(33.90916)\left(m_\mu c^2\right)(\gamma-1) + \left(m_\mu c^2\right)^2(\gamma-1)^2$$

Collecting terms, then solving for $(\gamma-1)$,

$$\gamma-1 = \frac{(33.90916)^2}{2\left(m_\mu c^2\right)^2 + 2(33.90916)m_\mu c^2} \qquad \text{Substituting } m_\mu c^2 = 105.65839\,MeV,$$

$\gamma-1 = 0.0390 \;\rightarrow\; \gamma = 1.0390$ so,

$$E_{k\mu} = 4.12\,MeV \quad \text{and} \quad p_\mu = \frac{1}{c}\left[(109.78)^2 - (105.66)^2\right]^{1/2} = 29.8 / MeV/c$$

$$E_v = 29.8\,MeV \quad \text{and} \quad p_v = 29.8\,MeV/c$$

(b)   If v_μ mass $= 250\,keV$, then $E_v^2 = \left(p_v c\right)^2$ and

$$E_{k\mu} + E_{kv} = 139.56755\,MeV - 105.65839\,MeV - 0.250\,MeV = 33.67916\,MeV$$

Solving as in (a) yields

$E_\mu = 109.78\,MeV,\; p_\mu = 29.8\,MeV/c,\; E_v = 29.8\,MeV,\; p_v = 29.8\,MeV/c$

2-53.   (a)   Energy and momentum are conserved.

Initial system: $E = Mc^2,\; p = 0$

     invariant mass: $\left(Mc^2\right)^2 = E^2 - (pc)^2 = \left(Mc^2\right)^2 + 0$

Final system:

     invariant mass: $\left(2mc^2\right)^2 = \left(Mc^2\right)^2 + 0$

For 1 particle (from symmetry)

$$\left(mc^2\right)^2 = \left(Mc^2/2\right)^2 - p^2c^2 = \left(Mc^2/2\right)^2 - (\gamma\,uc)^2$$

Rearranging,

$$1 = \left(\frac{Mc^2}{2mc^2}\right)^2 - (\gamma u/c)^2 \;\Rightarrow\; \gamma^2 = \left(\frac{Mc^2}{2mc^2}\right)^2 = \frac{1}{1-u^2/c^2}$$

Solving for u,

$$u = \left[1 - \left(\frac{2mc^2}{Mc^2}\right)^2\right]^{1/2} c$$

(b) Energy and momentum are conserved.

     Initial system: $E = 4mc^2$

        invariant mass: $\left(Mc^2\right)^2 = \left(4mc^2\right)^2 - (pc)^2$

(Problem 2-53 continued)

Final system:

invariant mass: $\left(2mc^2\right)^2 = \left(4mc^2\right)^2 - (pc)^2$

where $(pc)^2 = \left(4mc^2\right)^2 - \left(Mc^2\right)^2$

$$\frac{u}{c} = \frac{pc}{E} = \frac{\left[\left(4mc^2\right)^2 - \left(Mc^2\right)^2\right]^{1/2}}{4mc^2}$$

$$\left(\frac{u}{c}\right)^2 = \frac{\left(4mc^2\right)^2 - \left(Mc^2\right)^2}{\left(4mc^2\right)^2} = 1 - \left(\frac{Mc^2}{4mc^2}\right)^2$$

$$u = \left[1 - \left(\frac{Mc^2}{4mc^2}\right)^2\right]^{1/2} c$$

Chapter 3 – Quantization of Charge, Light, and Energy

3-1.   The radius of curvature is given by Equation 3-2.

$$R = \frac{mu}{qB} = m\left[\frac{2.5\times10^6\,m/s}{(1.60\times10^{-19}C)(0.40\,T)}\right] = m(3.91\times10^{25}\,m/s\cdot C\cdot T)$$

Substituting particle masses from Appendices A and D:

$$R(proton) = (1.67\times10^{-27}\,kg)(3.91\times10^{25}\,m/s\cdot C\cdot T) = 6.5\times10^{-2}\,m$$

$$R(electron) = (9.11\times10^{-31}\,kg)(3.91\times10^{25}\,m/s\cdot C\cdot T) = 3.6\times10^{-5}\,m$$

$$R(deuteron) = (3.34\times10^{-27}\,kg)(3.91\times10^{25}\,m/s\cdot C\cdot T) = 0.13\,m$$

$$R(H_2) = (3.35\times10^{-27}\,kg)(3.91\times10^{25}\,m/s\cdot C\cdot T) = 0.13\,m$$

$$R(helium) = (6.64\times10^{-27}\,kg)(3.91\times10^{25}\,m/s\cdot C\cdot T) = 0.26\,m$$

3-5.   (a) $R = \dfrac{mu}{qB} = \dfrac{\left[(2E_k/e)(e/m)\right]^{1/2}}{(e/m)(B)}$

$$= \frac{1}{B}\sqrt{\frac{2E_k/e}{e/m}} = \frac{1}{0.325\,T}\left[\frac{(2)(4.5\times10^4\,eV/e)}{1.76\times10^{11}\,kg}\right]^{1/2} = 2.2\times10^{-3}\,m = 2.2\,mm$$

(b) $frequency\ \ f = \dfrac{u}{2\pi R} = \dfrac{\sqrt{(2E_k/e)(e/m)}}{2\pi R}$

$$= \frac{\left[(2)(4.5\times10^4\,eV/e)(1.76\times10^{11}\,C/kg)\right]^{1/2}}{2\pi(2.2\times10^{-3}\,m)} = 9.1\times10^9\,Hz$$

$period\ \ T = 1/f = 1.1\times10^{-10}s$

3-9.   For the rise time to equal the field-free fall time, the net upward force must equal the weight.

$q\mathscr{E} - mg = mg\ \ \therefore\ \ \mathscr{E} = 2mg/q.$

3-13. Equation 3-10: $R = \sigma T^4$. Equation 3-12: $R = \dfrac{1}{4}cU$. From Example 3-5:

$$U = \left(8\pi^5 k^4 T^4\right) / \left(15 h^3 c^2\right)$$

$$\sigma = \frac{R}{T^4} = \frac{(1/4)cU}{T^4} = \frac{1}{4}c\left(8\pi^5 k^4 T^4\right)/\left(15 h^3 c^2 T^4\right)$$

$$= \frac{2\pi^5\left(1.38 \times 10^{-23}\,J/K\right)^4}{15\left(6.63 \times 10^{-34}\,J\cdot s\right)^3 \left(3.00 \times 10^8\,m/s\right)^2} = 5.67 \times 10^{-8}\,W/m^2 K^4$$

3-17. Equation 3-10: $R_1 = \sigma T_1^4 \quad R_2 = \sigma T_2^4 = \sigma\left(2T_1\right)^4 = 16\sigma T_1^4 = 16 R_1$

3-21. Equation 3-10:

$$R = \sigma T^4$$

$$P_{abs} = \left(1.36 \times 10^3\,W/m^2\right)\left(\pi R_E^2 m^2\right) \quad \text{where } R_E = \text{radius of Earth}$$

$$P_{emit} = \left(R\,W/m^2\right)\left(4\pi R_E^2\right) = \left(1.36 \times 10^3\,W/m^2\right)\left(\pi R_E^2 m^2\right)$$

$$R = \left(1.36 \times 10^3\,W/m^2\right)\left(\frac{\pi R_E^2}{4\pi R_E^2}\right) = \frac{1.36 \times 10^3}{4}\,\frac{W}{m^2} = \sigma T^4$$

$$T^4 = \frac{1.36 \times 10^3\,W/m^2}{4\left(5.67 \times 10^{-8}\,W/m^2 \cdot K^4\right)} \quad \therefore T = 278.3\,K = 5.3\,C$$

3-25. (a)

$$hf = hc/\lambda = 0.47\,eV.$$

$$\lambda_{max} = \frac{hc}{4.87\,eV} = \frac{\left(4.14 \times 10^{-15}\,eV\cdot s\right)\left(3.00 \times 10^8\,m/s\right)}{4.87\,eV} = 2.55 \times 10^{-7}\,m = 255\,nm$$

(b) It is the fraction of the total solar power with wavelengths less than 255 nm, i.e., the area

under the Planck curve (Figure 3-7) up to 255 nm divided by the total area. The latter

(Problem 3-25 continued)

is: $R = \sigma T^4 = (5.67 \times 10^{-8}\, W/m^2 \cdot K^4)(5800\, K)^4 = 6.42 \times 10^7\, W/m^2$. Approximating the former with

$u(\lambda)\Delta\lambda$ with $\lambda = 127\, nm$ and $\Delta\lambda = 255\, nm$:

$$[u(127\, nm)](255\, nm) = \left[\frac{8\pi h c (127 \times 10^{-9}\, m)^{-5}}{e^{hc/kT(127 \times 10^{-9})} - 1}\right](255 \times 10^{-9}\, m) = 1.23 \times 10^{-4}\, J/m^3$$

$$R(0\text{-}255\, nm) = \frac{c}{4}(1.23 \times 10^{-4}\, J/m^3) \;\Rightarrow\; \frac{R(0\text{-}255\, nm)}{R}$$

$$= \frac{(3.00 \times 10^8\, m/s)(1.23 \times 10^{-4}\, J/m^3)}{(4)(6.42 \times 10^7\, W/m^2)} \qquad fraction = 1.4 \times 10^{-4}$$

3-29.　(a) $E = \dfrac{hc}{\lambda} = \dfrac{1240\, eV \cdot nm}{0.1\, nm} = 1.24 \times 10^4\, eV$

(b) $E = \dfrac{hc}{\lambda} = \dfrac{1240\, eV \cdot nm}{1\, fm} \times \dfrac{10^6\, fm}{1\, nm} = 1.24 \times 10^9\, eV = 1.24\, GeV$

(c) $E = hf = (4.14 \times 10^{-15}\, eV \cdot s)(90.7 \times 10^6\, Hz) = 3.75 \times 10^{-7}\, eV$

3-33.　Equation 3-31: $\lambda_2 - \lambda_1 = \Delta\lambda = \dfrac{h}{mc}(1 - \cos\theta)$

$$\Delta\lambda = \frac{(6.63 \times 10^{-34}\, J \cdot s)(1 - \cos 135°)}{(9.11 \times 10^{-31}\, kg)(3.00 \times 10^8\, m/s)} = 4.14 \times 10^{-12}\, m = 4.14 \times 10^{-3}\, nm$$

$$\frac{\Delta\lambda}{\lambda_1} \times 100 = \frac{4.14 \times 10^{-3}\, nm}{0.0711\, nm} \times 100 = 5.8\%$$

3-37. $\Delta\lambda = \lambda_2 - \lambda_1 = \Delta\lambda = \dfrac{h}{mc}(1 - \cos\theta) = 0.01\,\lambda_1$ *Equation* 3-31

$$\lambda_1 = (100)\dfrac{h}{mc}(1 - \cos\theta) = (100)(0.00243\,nm)(1 - \cos 90°) = 0.243\,nm$$

3-41. (a) Compton wavelength $= \dfrac{h}{mc}$

electron: $\dfrac{h}{mc} = \dfrac{6.63 \times 10^{-34}\,J\cdot s}{(9.11 \times 10^{-31}\,kg)(3.00 \times 10^{8}\,m/s)} = 2.43 \times 10^{-12}\,m = 0.00243\,nm$

proton: $\dfrac{h}{mc} = \dfrac{6.63 \times 10^{-34}\,J\cdot s}{(1.67 \times 10^{-27}\,kg)(3.00 \times 10^{8}\,m/s)} = 1.32 \times 10^{-15}\,m = 1.32\,fm$

(b) $E = \dfrac{hc}{\lambda}$ (i) electron: $E = \dfrac{1240\,eV\cdot nm}{0.00243\,nm} = 5.10 \times 10^{5}\,eV = 0.510\,MeV$

(ii) proton: $E = \dfrac{1240\,eV\cdot nm}{1.32 \times 10^{-6}\,nm} = 9.39 \times 10^{8}\,eV = 939\,MeV$

3-45. Calculate $1/\lambda$ to be used in the graph.

$1/\lambda$ $(10^6/m)$	5.0	3.3	2.5	2.0	1.7
V_0 (V)	4.20	2.06	1.05	0.41	0.03

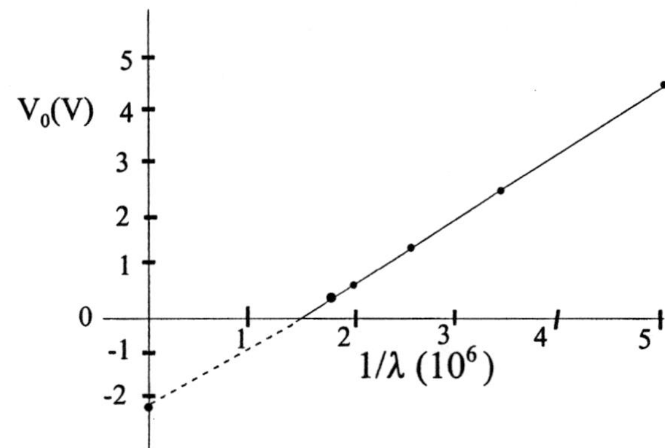

(Problem 3-45 continued)

(a) The intercept on the vertical axis is the work function ϕ. $\phi = 2.08$ eV

(b) The intercept on the horizontal axis corresponds to the threshold frequency.

$$\frac{1}{\lambda_t} = 1.65 \times 10^6 \,/m$$

$$f_t = \frac{c}{\lambda_t} = \left(3.00 \times 10^8 \, m/s\right)\left(1.65 \times 10^6 /m\right) = 4.95 \times 10^{14} \, Hz$$

(c) The slope of the graph is h/e. Using the vertical intercept and the largest experimental point,

$$\frac{h}{e} = \frac{1}{c}\frac{\Delta V_0}{\Delta(1/\lambda)} = \frac{4.20\,V - (-2.08\,V)}{\left(3.00 \times 10^8 \, m/s\right)\left(5.0 \times 10^6 /m - 0\right)} = 4.19 \times 10^{-15} \, eV/Hz$$

3-49.   Conservation of energy: $E_1 + mc^2 = E_2 + E_k + mc^2$ $\therefore$ $E_k = E_1 - E_2 = hf_1 - hf_2$

From Compton's equation, we have: $\lambda_2 - \lambda_1 = \dfrac{h}{mc}(1 - \cos\theta)$ thus $\dfrac{1}{f_2} - \dfrac{1}{f_1} = \dfrac{h}{mc^2}(1 - \cos\theta)$

$$\frac{1}{f_2} = \frac{1}{f_1} + \frac{h}{mc^2}(1 - \cos\theta) \quad \therefore \quad f_2 = \frac{f_1 mc^2}{mc^2 + hf_1(1 - \cos\theta)}$$

Substituting this expression for f_2 into the expression for E_k (and dropping the subscript on f_1):

$$E_k = hf - \frac{hfmc^2}{mc^2 + hf(1 - \cos\theta)} = \frac{hfmc^2 + (hf)^2(1 - \cos\theta) - hfmc^2}{mc^2 + hf(1 - \cos\theta)} = \frac{hf}{1 + \dfrac{mc^2}{[hf(1 - \cos\theta)]}}$$

E_k has its maximum value when the photon energy change is maximum, i.e., when $\theta = \pi$ so $\cos\theta = -1$. Then

$$E_k = \frac{hf}{1 + \dfrac{mc^2}{2hf}}$$

3-53.   (a) Equation 3-24: $u(\lambda) = \dfrac{8\pi hc\lambda^{-5}}{e^{hc/\lambda kT} - 1}$ Letting $C = 8\pi hc$ and $a = hc/kT$

gives $u(\lambda) = \dfrac{C\lambda^{-5}}{e^{a/\lambda} - 1}$

(b) $\dfrac{du}{d\lambda} = \dfrac{d}{d\lambda}\left[\dfrac{C\lambda_{-5}}{e^{a/\lambda} - 1}\right] = C\left[\dfrac{\lambda^{-5}(-1)e^{a/\lambda}(-a\lambda^{-2})}{(e^{a/\lambda} - 1)^2} - \dfrac{5\lambda^{-6}}{e^{a/\lambda} - 1}\right]$

$= \dfrac{C\lambda^{-6}}{(e^{a/\lambda} - 1)^2}\left[\dfrac{a}{\lambda}e^{a/\lambda} - 5(e^{a/\lambda} - 1)\right] = \dfrac{C\lambda^{-6}e^{a/\lambda}}{(e^{a/\lambda} - 1)^2}\left[\dfrac{a}{\lambda} - 5(1 - e^{a/\lambda})\right] = 0$

The maximum corresponds to the vanishing of the quantity in the brackets. Thus,
$5\lambda(1 - e^{-a/\lambda}) = a.$

(c) This equation is most efficiently solved by trial and error; i.e., guess at a value for a/λ in the expression $5\lambda(1 - e^{-a/\lambda}) = a$, solve for a better value of a/λ; substitute the new value to get an even better value, and so on. Repeat the process until the calculated value no longer changes. One succession of values is: 5, 4.966310, 4.965156, 4.965116, 4.965114, 4.965114. Further iterations repeat the same value (to seven digits), so we have $\dfrac{a}{\lambda_m} = 4.965114 = \dfrac{hc}{\lambda_m kT}$

(d) $\lambda_m T = \dfrac{hc}{(4.965114)k} = \dfrac{(6.63 \times 10^{-34}\,J\cdot s)(3.00 \times 10^8\,m/s)}{(4.965114)(1.38 \times 10^{-23}\,J/K)}.$

Therefore, $\lambda_m T = 2.898 \times 10^{-3}\,m\cdot K$ *Equation 3-11*

3-57.   (a) $E_k = 50\,keV$ and $\lambda_2 = \lambda_1 + 0.095\,nm$

$\dfrac{hc}{\lambda_1} + \dfrac{hc}{\lambda_2} = 5.0 \times 10^4\,eV$ $\therefore$ $\dfrac{1}{\lambda_1} + \dfrac{1}{\lambda_1 + 0.095} = \dfrac{5.0 \times 10^4\,eV}{hc}$

$\therefore$ $\dfrac{2\lambda_1 + 0.095}{\lambda_1^2 + 0.095\,\lambda_1} = \dfrac{5.0 \times 10^4\,eV}{hc}$

$\lambda_1^2 + \left(0.095\,nm - \dfrac{2hc}{5 \times 10^4\,eV}\right)\lambda_1 - \dfrac{(0.095\,nm)hc}{5 \times 10^4\,eV} = 0$

$\therefore$ $\lambda_1^2 + 0.04532\,\lambda_1 - 2.35 \times 10^{-3} = 0$

(Problem 3-57 continued)

Applying the quadratic formula,

$$\lambda_1 = \frac{-0.04532 \pm \left[(0.04532)^2 + 4\left(2.3598 \times 10^{-3}\right)\right]^{1/2}}{2}$$

$$\lambda_1 = 0.06189 \, nm \quad and \quad \lambda_2 = 0.08139 \, nm$$

(b) $E_1 = \dfrac{hc}{\lambda_1} = \dfrac{1240 \, eV \cdot nm}{0.06189 \, nm} = 20.04 \, keV$

Chapter 4 – The Nuclear Atom

4-1. $\dfrac{1}{\lambda_{mn}} = R\left(\dfrac{1}{m^2} - \dfrac{1}{n^2}\right)$ where $R = 1.097 \times 10^7 m^{-1}$ (Equation 4-2)

The Lyman series ends on m = 1, the Balmer series on m = 2, and the Paschen series on m = 3. The series limits all have n = ∞, so $\dfrac{1}{n} = 0$

$$\frac{1}{\lambda_L} = R\left(\frac{1}{1^2}\right) = 1.097 \times 10^7 \, m^{-1}$$

$$\lambda_L(limit) = 1.097 \times 10^7 m^{-1} = 91.16 \times 10^{-9} m = 91.16 \, nm$$

$$\frac{1}{\lambda_B} = R\left(\frac{1}{2^2}\right) = 1.097 \times 10^7 \, m^{-1}/4$$

$$\lambda_B(limit) = 4/1.097 \times 10^7 m^{-1} = 3.646 \times 10^{-7} m = 364.6 \, nm$$

$$\frac{1}{\lambda_P} = R\left(\frac{1}{3^2}\right) = 1.097 \times 10^7 \, m^{-1}/9$$

$$\lambda_P(limit) = 9/1.097 \times 10^7 m^{-1} = 8.204 \times 10^{-7} m = 820.4 \, nm$$

4-5. None of these lines are in the Paschen series, whose limit is 820.4 nm (see Problem 4-1)
and whose first line is given by: $\dfrac{1}{\lambda_{34}} = R\left(\dfrac{1}{3^2} - \dfrac{1}{4^2}\right) \rightarrow \lambda_{34} = 1875 \, nm$. Also, none are in

the Brackett series, whose longest wavelength line is 4052 nm (see Problem 4-4).
The Pfund series has $m = 5$. Its first three (i.e., longest wavelength) lines have $n = 6, 7$,
and 8.

$$\frac{1}{\lambda_{56}} = R\left(\frac{1}{5^2} - \frac{1}{6^2}\right) = 1.341 \times 10^5 \, m^{-1}$$

$\lambda_{56} = 1/1.341 \times 10^5 \, m^{-1} = 7.458 \times 10^{-6} m = 7458 \, nm$. Similarly,

(Problem 4-5 continued)

$$\lambda_{57} = 1/2.155 \times 10^5 \, m^{-1} = 4.653 \times 10^{-6} \, m = 4653 \, nm$$

$$\lambda_{58} = 1/2.674 \times 10^5 \, m^{-1} = 3.740 \times 10^{-6} \, m = 3740 \, nm$$

Thus, the line at 4103 nm is not a hydrogen spectral line.

4-9.
$$r_d = \frac{kq_\alpha Q}{\frac{1}{2} m_\alpha v^2} = \frac{ke^2 \cdot 2 \cdot 79}{E_{k\alpha}} \qquad (\text{Equation } 4\text{-}11)$$

For $E_{k\alpha} = 5.0 \, MeV$: $r_d = \dfrac{(1.44 \, MeV \cdot fm)(2)(79)}{5.0 \, MeV} = 45.5 \, fm$

For $E_{k\alpha} = 7.7 \, MeV$ $\quad r_d = 29.5 \, fm$

For $E_{k\alpha} = 12 \, MeV$ $\quad r_d = 19.0 \, fm$

4-13. (a) $r_n = \dfrac{n^2 a_0}{Z}$ Equation 4-18

$$r_6 = \frac{6^2 (0.053 \, nm)}{1} = 1.91 \, nm$$

(b) $r_6(He^+) = \dfrac{6^2 (0.053 \, nm)}{2} = 0.95 \, nm$

4-17. (a) $\lambda = 410.7 \, nm$ is in the visible region of the spectrum, so this is a transition ending on

$n = 2$ (see Figure 4-16).

(Problem 4-17 continued)

$$\frac{hc}{\lambda} = E_n - E_2 = 13.6\,eV\left(\frac{1}{2^2} - \frac{1}{n^2}\right)$$

$$\frac{1240\,eV\cdot nm}{410.7\,nm} = 13.6\,eV\left(\frac{1}{4} - \frac{1}{n^2}\right)$$

$$\frac{1}{n^2} = \frac{1}{4} - \frac{1240\,eV\cdot nm}{410.7\,nm\,(13.6\,eV)} = 0.0280$$

$$n = \left(\frac{1}{0.0280}\right)^{1/2} = 6$$

(b)   This line is in the Balmer series.

4-21.

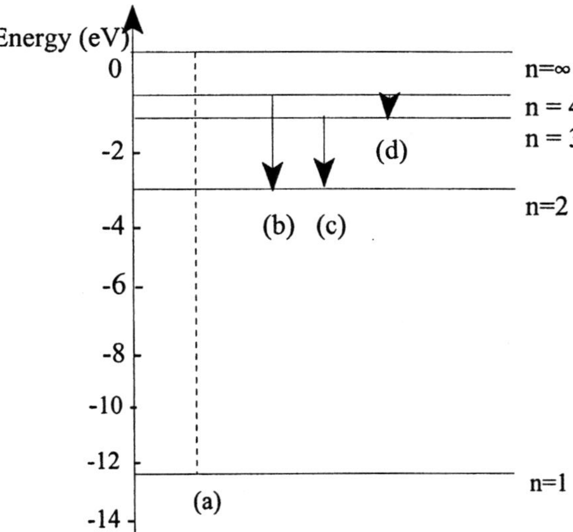

(a) Lyman limit (b) H_β line (c) H_α line (d) 1st line of Paschen series

4-25.   (a) The result of the Bohr orbits are given by (Equation 4-18) $r = n^2 a_0/Z$ where $a_0 =$
0.0529 nm and $Z = 1$ for hydrogen. For $n = 600$, $r = (600)^2(0.0529\text{ nm}) = 1.90 \times 10^4$
nm = 19.0 μm.   This is about the size of a tiny grain of sand.
(b) The electron's speed in a Bohr orbit is given by

(Problem 4-25 continued)

$v^2 = ke^2/mr$ with $Z = 1$

Substituting r for the $n = 600$ orbit from (a), then taking the square root,

$$v^2 = \left(8.99 \times 10^9 \, N \cdot m^2\right)\left(1.60 \times 10^{-19} \, C\right)^2 / \left(9.11 \times 10^{-31} \, kg\right)\left(19.0 \times 10^{-6} \, \mu m\right)$$

$$v^2 = 1.33 \times 10^7 \, m^2/s^2$$

$$v = 3.65 \times 10^3 \, m/s$$

For comparison, in the $n = 1$ orbit, v is about 2×10^6 m/s.

4-29. $r_n = \dfrac{n^2 a_0}{Z}$ (Equation 4-18)

The $n = 1$ electrons "see" a nuclear charge of approximately $Z-1$, or 78 for Au.

$$r_1 \approx 0.0529 \, nm / 78 = 6.8 \times 10^{-4} \, nm \approx 6.8 \times 10^{-4} \, nm \left(10^{-9} \, m/nm\right)\left(10^{15} \, fm/m\right) = 680 \, fm$$

or about 100 times the radius of the Au nucleus.

4-33.

Element	Al	Ar	Sc	Fe	Ge	Kr	Zr	Ba
Z	13	18	21	26	32	36	40	56
E (keV)	1.56	3.19	4.46	7.06	10.98	14.10	17066	36.35
$f^{1/2}$ (10^8 Hz$^{1/2}$)	6.14	8.77	10.37	13.05	16.28	18.45	20.64	29.62

(Problem 4-33 continued)

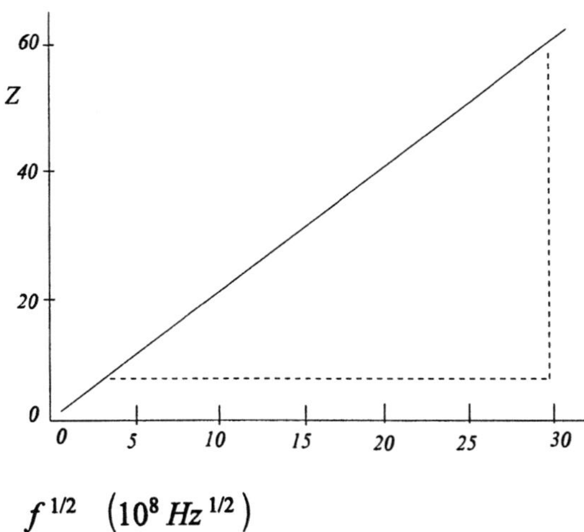

$$\text{slope} = \frac{58 - 10}{(30 - 4.8) \times 10^8} = 1.90 \times 10^{-8}\ Hz^{-1/2}$$

$$\text{slope (Fig. 4-18)} = \frac{30 - 13}{(15 - 7) \times 10^8} = 2.13 \times 10^{-8}\ Hz^{-1/2}$$

The two values are in good agreement.

4-37.   Using the results from Problem 4-24, the energy of the positronium Lyman α line is
$\Delta E = E_2 - E_1 = -1.701\,eV - (-6.804\,eV) = 5.10\ eV$. The first Franck-Hertz current decrease
would occur at 5.10 V, the second at 10.2 V.

4-41.   (a) $\displaystyle i = q\,f_{rev} = e\,\frac{Z^2 m k^2 e^4}{2\pi\hbar^3 n^3}$ (from Equation 4-28)

$$= e\,\frac{m c^2 (ke^2)^2 (1)^2}{2\pi\hbar(\hbar c)^2 (1)^3} = \frac{ec}{(h/mc)}\left(\frac{ke^2}{\hbar c}\right)^2 = \frac{ec\,\alpha^2}{\lambda_c}$$

$$= \frac{(1.602 \times 10^{-19}\,C)(3.00 \times 10^{17}\,nm/s)}{0.00243\,nm}\left(\frac{1}{137}\right)^2 = 1.054 \times 10^{-3}\ A$$

(Problem 4-41 continued)

(b)
$$\mu = i A = i \pi a_o^2 = \left(\frac{emk^2e^4}{2\pi\hbar^3}\right) \pi \left(\frac{\hbar^2}{mke^2}\right) = \frac{e\hbar}{2m}$$

$$= \frac{(1.602\times10^{-19}\,C)(1.055\times10^{-34}\,J\cdot s)}{2(9.11\times10^{-31}\,kg)} = 9.28\times10^{-24}\,A\cdot m^2$$

or

$$= (1.054\times10^{-3}\,A)\,\pi\,(0.529\times10^{-10}\,m)^2 = 9.27\times10^{-24}\,A\cdot m^2$$

4-45. (a) $E_n = -E_0 Z^2 / n^2$ (Equation 4-20)

For Li^{++}, $Z = 3$ and $E_n = -13.6\,eV(9)/n^2 = -122.4/n^2\,eV$

The first three Li^{++} levels that have the same (nearly) energy as H are:

n = 3, $E_3 = -13.6$ eV   n = 4, $E_6 = -3.4$ eV   n = 9, $E_9 = -1.51$ eV

Lyman α corresponds to the $n = 6 \rightarrow n = 3\,Li^{++}$ transition. Lyman β corresponds to the $n = 9 \rightarrow n = 3\,Li^{++}$ transition.

(b) $R(H) = R_\infty(1/(1 + 0.511\,MeV/938.8\,MeV)) = 1.096776\times10^7\,m^{-1}$

$R(Li) = R_\infty(1/(1 + 0.511\,MeV/6535\,MeV)) = 1.097287\times10^7\,m^{-1}$

For Lyman α:

$$\frac{1}{\lambda} = R(H)\left(1 - \frac{1}{2^2}\right) = 1.096776\times10^7\,m^{-1}(10^{-9}\,m/nm)(3/4) \rightarrow \lambda = 121.568\,nm$$

For Li^{++} equivalent:

$$\frac{1}{\lambda} = R(Li)\left(\frac{1}{3^2} - \frac{1}{6^2}\right) Z^2 = 1.097287\times10^7\,m^{-1}(10^{-9}\,m/nm)\left(\frac{1}{9} - \frac{1}{36}\right)(3)^2$$

$\lambda = 121.512\,nm$ $\Delta\lambda = 0.056\,nm$

4-49. (a) $b = R\,\sin\beta = R\,\sin\left(\frac{180° - \theta}{2}\right) = R\,\cos\frac{\theta}{2}$

(b) Scattering through an angle larger than θ corresponds to an impact parameter smaller than b. Thus, the shot must hit within a circle of radius b and area πb^2. The rate at which this occurs is $I_o \pi b^2 = I_o R^2 \cos^2\frac{\theta}{2}$

(Problem 4-49 continued)

(c) $\quad \sigma = \pi b_o^2 = \pi \left(R \cos\dfrac{0}{2} \right)^2 = \pi R^2$

(d) $\quad$ An α particle with an arbitrarily large impact parameter still feels a force and is scattered.

4-53. $\quad \dfrac{kZe^2}{r} = \dfrac{mv^2}{r} \;\rightarrow\; \dfrac{kZe^2}{r^2} = \dfrac{(\gamma mv)^2}{mr} \quad$ (from Equation 4-12)

$$\gamma v = \left(\dfrac{kZe^2}{mr} \right)^{1/2} = \dfrac{v}{\sqrt{1-\beta^2}}$$

$$\dfrac{c^2\beta^2}{1-\beta^2} = \left(\dfrac{kZe^2}{mr} \right) \quad \text{Therefore,} \quad \beta^2\left[c^2 + \left(\dfrac{kZe^2}{mr} \right) \right] = \left(\dfrac{kZe^2}{mr} \right)$$

$$\beta^2 \approx \dfrac{1}{c^2}\left(\dfrac{kZe^2}{ma_o} \right) \;\rightarrow\; \beta = 0.0075\,Z^{1/2} \;\rightarrow\; v = 0.0075\,cZ^{1/2} = 2.25 \times 10^6\,m/s \times Z^{1/2}$$

$$E = KE - kZe^2/r = mc^2(\gamma - 1) - \dfrac{kZe^2}{r} = mc^2\left[\dfrac{1}{\sqrt{1-\beta^2}} - 1 \right] - \dfrac{kZe^2}{r}$$

and substituting $\beta = 0.0075$ and $r = a_0$

$$E = 511 \times 10^3\,eV\left[\dfrac{1}{\sqrt{1-(0.0075)^2}} - 1 \right] - 28.8\,ZeV$$

$$= 14.4\,eV - 28.8\,ZeV = -14.4\,Z\,eV$$

4-57. $\quad$ Refer to Figure 4-16. All possible transitions starting at $n = 5$ occur.

$n = 5$ to $n = 4, 3, 2, 1$

$n = 4$ to $n = 3, 2, 1$

$n = 3$ to $n = 2, 1$

$n = 2$ to $n = 1$

Thus, there are 10 different photon energies emitted.

(Problem 4-57 continued)

n_i	n_f	fraction	no. of photons
5	4	1/4	125
5	3	1/4	125
5	2	1/4	125
5	1	1/4	125
4	3	$1/4 \times 1/3$	42
4	2	$1/4 \times 1/3$	42
4	1	$1/4 \times 1/3$	42
3	2	$1/2\left[1/4 + 1/4(1/3)\right]$	83
3	1	$1/2\left[1/4 + 1/4(1/3)\right]$	83
2	1	$\left[1/2\left(1/4 + 1/4\right)(1/3)) + 1/4(1/3) + 1/4\right]$	250

$$\text{Total} = 1{,}042$$

Note that the number of electrons arriving at the $n = 1$ level $(125 + 42 + 83 + 250)$ is 500, as it should be.

Chapter 5 – The Wavelike Properties of Particles

5-1.   (a) $\lambda = \dfrac{h}{p} = \dfrac{h}{mv} = \dfrac{(6.63 \times 10^{-34}\,J\cdot s)(3.16 \times 10^{7}\,s/y)}{(10^{-3}\,kg)(1\,m/y)} = 2.1 \times 10^{-23}\,m$

(b) $v = \dfrac{h}{m\lambda} = \dfrac{6.63 \times 10^{-34}\,J\cdot s}{(10^{-3}\,kg)(10^{-2}\,m)} = 6.6 \times 10^{-29}\,m/s = 2.1 \times 10^{-21}\,m/y$

5-5. $\lambda = h/p = h/\sqrt{2mE_k} = hc/\left[2mc^2(1.5\,kT)\right]^{1/2}$ (from Equation 5-2)

Mass of N_2 molecule $= 2 \times 14.0031\,u(931.5\,MeV/uc^2) = 2.609 \times 10^4\,MeV/c^2 = 2.609 \times 10^{10}\,eV/c^2$

$$\lambda = \frac{1240\,eV\cdot nm}{\left[(2)(2.609 \times 10^{10}\,eV)(1.5)(8.617 \times 10^{-5}\,eV/K)(300\,K)\right]^{1/2}} = 0.0276\,nm$$

5-9. $E_k = mc^2(\gamma - 1)$ $p = \gamma mu$

(a) $E_k = 2\,GeV$ $mc^2 = 0.938\,GeV$

$\gamma - 1 = E_k/mc^2 = 2\,GeV/0.938\,GeV = 2.132$ Thus, $\gamma = 3.132$

Because, $\gamma = 1/\sqrt{1-(u/c)^2}$ where $u/c = 0.948$

$$\lambda = \frac{h}{p} = \frac{h}{\gamma mc(u/c)} = \frac{hc}{\gamma mc^2(u/c)}$$

$$= \frac{1240\,eV\cdot nm}{(3.132)(938 \times 10^6\,eV)(0.948)} = 4.45 \times 10^{-7}\,nm = 0.445\,fm$$

(b) $E_k = 200\,GeV$

$\gamma - 1 = E_k/mc^2 = 200\,GeV/0.938\,GeV = 213$. Thus, $\gamma = 214$ and $u/c = 0.9999$

$$\lambda = \frac{1240\,MeV\cdot fm}{(214)(938\,MeV)(0.9999)} = 6.18 \times 10^{-3}\,fm$$

5-13.

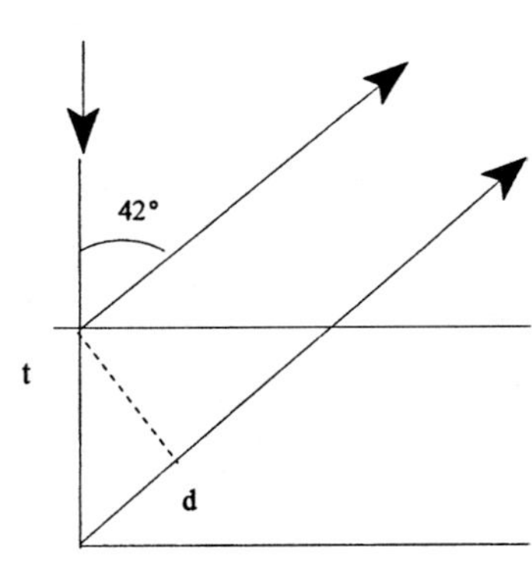

$d = t\cos 42°$

$n\lambda = t + d = t(1 + \cos 42°) = 0.30\ nm\,(1 + \cos 42°)$

For the first maximum n = 1, so $\lambda = 0.523\ nm$

$$\lambda = \frac{h}{p} = \frac{h}{\sqrt{2mE_k}} \quad\rightarrow\quad E_k = \frac{h^2}{2m\lambda_2} = \frac{(hc)^2}{2mc^2\lambda^2}$$

$$E_k = \frac{(1240\ eV\cdot nm)^2}{2(939.6\times 10^6\ eV)(0.523\ nm)^2} = 3.0\times 10^{-3}\ eV$$

5-17. (a) $\quad y = y_1 + y_2$

$$= 0.002\,m\,\cos(8.0x\,/m - 400t\,/s) + 0.002\,m\,\cos(7.6x\,/m - 380t\,/s)$$

$$= 2(0.002\,m)\cos\left[\frac{1}{2}(8.0x\,/m - 7.6x\,/m) - \frac{1}{2}(400t\,/s - 380t\,/s)\right]$$

$$\times \cos\left[\frac{1}{2}(8.0x\,/m + 7.6x\,/m) - \frac{1}{2}(400t\,/s + 380t\,/s)\right]$$

$$= 0.004\,m\,\cos(0.2x\,/m - 10t\,/s) \times \cos(7.8x\,/m - 390t\,/s)$$

(b) $\quad v = \dfrac{\bar{\omega}}{\bar{k}} = \dfrac{390\,/s}{7.8\,/m} = 50\ m/s$

(c) $\quad v_g = \dfrac{\Delta\omega}{\Delta k} = \dfrac{20\,/s}{0.4\,/m} = 50\ m/s$

(d) Successive zeros of the envelope requires that $0.2\,\Delta x\,/m = \pi$, thus $\Delta x = \dfrac{\pi}{0.2} = 5\pi\,m$ with

$\quad \Delta k = k_1 - k_2 = 0.4\,m^{-1}$ and $\Delta x = \dfrac{2\pi}{\Delta k}$.

5-21. $\quad \Delta\omega\,\Delta t \approx 1 \quad\rightarrow\quad (2\pi\Delta f)\,\Delta t = 1 \quad$ Thus, $\Delta t \approx 1/(2\pi\times 5000) = 3.2\times 10^{-5}\,s$

5-25.　(a)　At $x = 0$:　$P\,dx = |\psi(0,0)|^2\,dx = |A\,e^0|^2\,dx = A^2\,dx$

　　　(b)　At $x = \sigma$:　$P\,dx = |A\,e^{-\sigma^2/4\sigma^2}|^2\,dx = |A\,e^{-1/4}|^2\,dx = 0.61\,A^2\,dx$

　　　(c)　At $x = 2\sigma$:　$P\,dx = |A\,e^{-4\sigma^2/4\sigma^2}|^2\,dx = |A\,e^{-1}|^2\,dx = 0.14\,A^2\,dx$

　　　(d)　The electron will most likely be found at x = 0, where $P\,dx$ is largest.

5-29.　$\Delta E\,\Delta t \approx \hbar \;\rightarrow\; \Delta E \approx \hbar/\Delta t = \dfrac{6.58 \times 10^{-16}\,eV{\cdot}s}{3.823\,d\left(8.64 \times 10^4\,s/d\right)} \approx 1.99 \times 10^{-21}\,eV$

The energy uncertainty of the excited state is ΔE, so the α energy can be no sharper than ΔE.

5-33.　(a)　For ^{48}Ti: $\Delta E_u(upper\ state) \approx \hbar/\Delta t = \dfrac{1.055 \times 10^{-34}\,J{\cdot}s}{1.4 \times 10^{-14}\,s\left(1.60 \times 10^{-13}\,J/MeV\right)} \approx 4.71 \times 10^{-10}\,MeV$

　　　　$\Delta E_L(lower\ state) \approx \hbar/\Delta t = \dfrac{1.055 \times 10^{-34}\,J{\cdot}s}{3.0 \times 10^{-12}\,s\left(1.60 \times 10^{-13}\,J/MeV\right)} \approx 2.20 \times 10^{-10}\,MeV$

　　　　$\Delta E\,(total) = \Delta E_u + \Delta E_L = 6.91 \times 10^{-10}\,MeV$

　　　　$\dfrac{\Delta E_T}{E} = \dfrac{6.91 \times 10^{-10}\,MeV}{1.312\,MeV} = 5.3 \times 10^{-10}$

　　　(b)　For Hα: $\Delta E_u \approx \dfrac{1.055 \times 10^{-34}\,J{\cdot}s}{10^{-8}\,s\left(1.60 \times 10^{-19}\,J/eV\right)} \approx 6.59 \times 10^{-8}\,eV$

　　　　and $\Delta E_L \approx 6.59 \times 10^{-8}\,eV$ also.

　　　　$\Delta E_T = 1.32 \times 10^{-7}\,eV$ is the uncertainty in the Hα transition energy of 1.9 eV.

5-37. $E = hf \rightarrow \Delta E = h\Delta f$

$\Delta E \Delta t \approx h \rightarrow \Delta f \Delta t \approx 1$ where $\Delta t = 0.85$ ms

$\Delta f = 1/0.85$ ms $= 1.$

For $\lambda = 0.01$ nm $\qquad f = c/\lambda = \dfrac{3.00 \times 10^8 \ m/s \times 10^9 \ nm/s}{0.01 \ nm} \ 18 \times 10^9$ Hz

$f = 3.00 \times 10^{19}$ Hz

$\dfrac{\Delta f}{f} = \dfrac{1.18 \times 10^9 \ Hz}{3.00 \times 10^{19} \ Hz} = 3.9 \times 10^{-11}$

5-41. $E = \overline{E} = \dfrac{\overline{p^2}}{2m} + \dfrac{1}{2}m\omega^2 \overline{x^2} = \dfrac{(\Delta p)^2}{2m} + \dfrac{1}{2}m\omega^2(\Delta x)^2.$ Substitute $\Delta p = \dfrac{\hbar}{2\Delta x}$

$E = \dfrac{\hbar^2}{8m(\Delta x)^2} + \dfrac{1}{2}m\omega^2(\Delta x)^2.$ To minimize E, set $dE/d(\Delta x) = 0$

$0 = \dfrac{dE}{d(\Delta x)} = \dfrac{\hbar^2}{8m}\dfrac{-2}{(\Delta x)^3} + m\omega^2\Delta x = \dfrac{m\omega^2}{(\Delta x)^3}\left[-\dfrac{\hbar^2}{4m^2\omega^2} + (\Delta x)^4\right]$

$\therefore \quad (\Delta x)^2 = \dfrac{\hbar}{2m\omega}$

$E_{min} = \dfrac{\hbar^2}{8m} \cdot \dfrac{2m\omega}{\hbar} + \dfrac{m\omega^2}{2}\dfrac{\hbar}{2m\omega} = \dfrac{1}{4}\hbar\omega + \dfrac{1}{4}\hbar\omega = \dfrac{1}{2}\hbar\omega$

$\omega = \sqrt{\dfrac{K}{m}} = \sqrt{\dfrac{(1\ N/m)}{10^{-2}\ kg}} = 10/s$

$E_{min} = \dfrac{1}{2}\hbar\omega = \dfrac{1}{2}(1.055 \times 10^{-34}\ J\cdot s)(10\ /s) = 5.27 \times 10^{-34}\ J$

5-45. $\Delta p = m\,\Delta v = m(0.0001)(500\,m/s) = 0.05\,m$

For proton: $\Delta x\,\Delta p \approx \hbar$

$$\Delta x \approx \hbar/\Delta p = \left(6.58\times 10^{-16}\,eV\cdot s\right)/(0.05\,m/s)\left(938\times 10^{6}\,eV\right)$$

$$\approx 1.40\times 10^{-23}\,m = 1.40\times 10^{-8}\,fm$$

For bullet: $\Delta x \approx = \left(1.055\times 10^{-34}\,J\cdot s\right)/(0.05\,m/s)\left(10\times 10^{-3}\,kg\right) \approx 2.1\times 10^{-31}\,m$

5-49.

$$hf = \gamma\,mc^2 \;\rightarrow\; \gamma = \frac{hf}{mc^2} = \frac{1}{\sqrt{1-v^2/c^2}}$$

$$1-v^2/c^2 = \left(\frac{mc^2}{hf}\right)^2$$

$$\frac{v}{c} = \left[1-\left(\frac{mc^2}{hf}\right)^2\right]^{1/2}$$

Expanding the right side, assuming $mc^2 << hf$,

$$\frac{v}{c} = 1-\frac{1}{2}\left(\frac{mc^2}{hf}\right)^2 - \frac{1}{8}\left(\frac{mc^2}{hf}\right)^4 +\cdots \quad \text{and neglecting all but the first two terms,}$$

$$\frac{v}{c} = 1-\frac{1}{2}\left(\frac{mc^2}{hf}\right)^2 \quad \text{Solving this for } m \text{ and inserting deBroglie's assumptions that}$$

$$\frac{v}{c} \geq 0.99 \text{ and } \lambda = 30\,m, \; m \text{ is then :}$$

$$m = \frac{\left[(1-0.99)2\right]^{1/2}\left(6.63\times 10^{-34}\,J\cdot s\right)}{\left(3.00\times 10^{8}\,m/s\right)(30\,m)} = 1.04\times 10^{-44}\,kg$$

Chapter 6 – The Schrödinger Equation

6-1. $\dfrac{d\Psi}{dx} = kAe^{kx-\omega t} = k\Psi$ and $\dfrac{d^2\Psi}{dx^2} = k^2\Psi$

Also, $\dfrac{d\Psi}{dt} = -\omega\Psi$. The Schrödinger equation is then, with these substitutions,

$-\hbar^2 k^2 \Psi / 2m + V\Psi = -i\hbar\omega\Psi$. Because the left side is real and the right side is a pure imaginary number, the proposed Ψ does not satisfy Schrödinger's equation.

6-5. (a) $\Psi(x,t) = A\sin(kx - \omega t)$

$$\frac{\partial\Psi}{\partial t} = -\omega A\cos(kx - \omega t)$$

$$i\hbar\frac{\partial\Psi}{\partial t} = -i\hbar\omega A\cos(kx - \omega t)$$

$$\frac{\partial^2\Psi}{\partial x^2} = -k^2 A\sin(kx - \omega t)$$

$$\frac{-\hbar^2}{2m}\frac{\partial^2\Psi}{\partial x^2} = \frac{-\hbar k^2 A}{2m}\sin(kx - \omega t) \neq i\hbar\frac{\partial\Psi}{\partial t}$$

(b) $\Psi(x,t) = A\cos(kx - \omega t) + iA\sin(kx - \omega t)$

$$i\hbar\frac{\partial\Psi}{\partial t} = i\hbar\omega A(kx - \omega t) - i^2\hbar\omega A\cos(kx - \omega t)$$

$$= \hbar\omega A\cos(kx - \omega t) + i\hbar\omega A\sin(kx - \omega t)$$

$$-\frac{\hbar^2}{2m}\frac{\partial^2\Psi}{\partial x^2} = \frac{\hbar^2 k^2 A}{2m}\cos(kx - \omega t) + \frac{\hbar^2 i k^2 A}{2m}\sin(kx - \omega t)$$

$$= \frac{\hbar^2 k^2}{2m}\left[A\cos(kx - \omega t) + iA\sin(kx - \omega t)\right]$$

$$= i\hbar\frac{\partial\Psi}{\partial t} \quad if \quad \frac{\hbar^2 k^2}{2m} = \hbar\omega \quad \text{it does.} \quad \text{(Equation 6-5 with } V = 0)$$

6-9.   (a) The ground state of an infinite well is $E_1 = h^2/8mL^2 = (hc)^2/8mc^2L^2$

$$\text{For } m = m_p, \, L = 0.1 \, nm: \quad E_1 = \frac{(1240 \, MeV \cdot fm)^2}{8(938.3 \times 10^6 \, eV)(0.1 \, nm)^2} = 0.021 \, eV$$

(b) For $m = m_p$, $L = 1 \, fm$: $\quad E_1 = \dfrac{(1240 \, MeV \cdot fm)^2}{8(938.3 \times 10^6 \, eV)(1 \, fm)^2} = 205 \, MeV$

6-13.   (a) $\Delta x = 0.0001 L = (0.0001)(10^{-2} m) = 10^{-6} m$

$$\Delta p = 0.0001 p = (0.0001)(10^{-9} kg)(10^{-3} m/s) = 10^{-16} \, kg \cdot m/s$$

(b) $\dfrac{\Delta x \, \Delta p}{\hbar} = \dfrac{(10^{-6} m)(10^{-16} kg \cdot m/s)}{1.055 \times 10^{-34} \, J \cdot s} = 9 \times 10^{11}$

6-17.   This is an infinite square well with $L = 10 \, cm$.

$$E_n = \frac{h^2 n^2}{8mL^2} = \frac{1}{2} mv^2 = \frac{(2.0 \times 10^{-3} kg)(20 \, nm)^2}{2(3.16 \times 10^7 s)^2}$$

$$n^2 = \frac{8(2.0 \times 10^{-3} kg)^2 (20 \times 10^{-9} m)^2 (0.1 \, m)^2}{2(3.16 \times 10^7 s)^2 (6.63 \times 10^{-34} J \cdot s)^2}$$

$$n = \frac{2(2.0 \times 10^{-3} kg)(20 \times 10^{-9} m)(0.1 \, m)}{3.16 \times 10^7 s \, (6.63 \times 10^{-34} J \cdot s)} = 3.8 \times 10^{14}$$

6-21.
$$\psi_n(x) = \sqrt{\frac{2}{L}} \, \sin\frac{n\pi x}{L}$$

To show that

$$\int_0^L \sin\left(\frac{n\pi x}{L}\right) \sin\left(\frac{n\pi x}{L}\right) dx = 0$$

Using the trig identity $2 \, sinA \, sinB = cos(A-B) - cos(A+B)$, the integrand becomes

(Problem 6-21 continued)

$$\frac{1}{2}\{\cos[(n-m)\pi x/L] - \cos[(n+m)\pi x/L]\}$$

The integral of the first term is

$$\frac{L}{\pi}\frac{\sin(n-m)\pi x/L}{(n-m)} \quad \text{and similarly for the second term with } (n+m) \text{ replacing}$$

$(n-m)$. Since n and m are integers and $n\neq m$, the sines both vanish at the limits $x=0$ and $x=L$.

$$\therefore \int_0^L \sin\left(\frac{n\pi x}{L}\right)\sin\left(\frac{m\pi x}{L}\right)dx = 0 \qquad \text{for } n\neq m.$$

6-25.   For $V_2 > E > V_1$: x_1 is where $V=0 \rightarrow V_1$ and x_2 and x_2 is where $V_1 \rightarrow V_2$

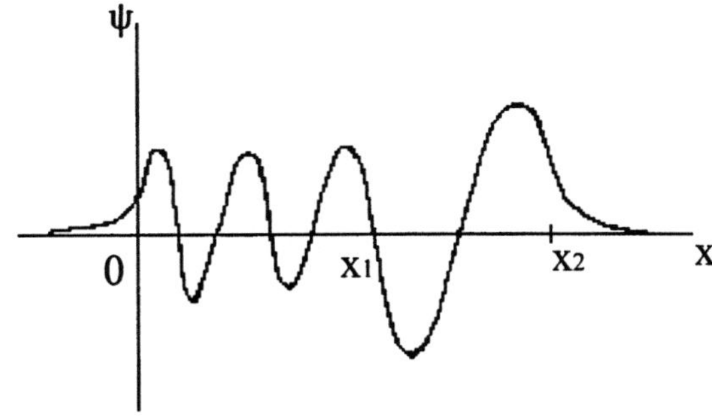

From $-\infty$ to 0 and x_2 to $+\infty$: ψ is exponential

0 to x_1 : ψ is oscillatory; E_k is large so p is large and λ is small; amplitude is small because
    E_k, hence v is large.
x_1 to x_2: ψ is oscillatory; E_k is small so p is small and λ is large; amplitude is large because
    E_k, hence v is small.

6-29. (a) Classically, the particle is equally likely to be found anywhere in the box, so $P(x) =$ constant. In addition, $\int_0^L P(x)\,dx = 1$ so $P(x) = 1/L$.

(b) $\langle x \rangle = \int_0^L (x/L)\,dx = L/2$ and $\langle x^2 \rangle = \int_0^L (x^2/L)\,dx = L^2/3$

6-33. $\dfrac{p^2}{2m} + \dfrac{1}{2}m\omega^2 x^2 = (n+1/2)\hbar\omega$. For the ground state (n = 0),

$$\langle x^2 \rangle = \frac{2}{m\,\omega^2}\left(\hbar\omega/2 - p^2/2m\right) \quad \text{and} \quad \langle x \rangle^2 = \left(\frac{\hbar}{m\omega} - \frac{p^2}{m^2\omega^2}\right) = \hbar/2m\omega \ \text{(See Problem 6-32)}$$

$$\frac{\hbar}{m\omega}\left(1 - \frac{p^2}{m\hbar\omega}\right) = \frac{\hbar}{2m\omega} \quad \text{or} \quad \left(1 - \frac{p^2}{m\hbar\omega}\right) = \frac{1}{2} \quad \rightarrow \quad \langle p^2 \rangle = \frac{1}{2}m\hbar\omega$$

6-37. (a) $\Delta x\,\Delta p \approx \hbar \quad \rightarrow \quad \Delta p \approx \hbar/\Delta x \approx \hbar/2A$

(b) $E_k = p^2/2m \approx (\hbar/2A)^2/2m \approx \hbar^2/8mA^2$

$$(1)\ E_0 = \frac{1}{2}m\omega^2 A^2\left(\frac{\hbar^2}{\hbar^2}\right)\left(\frac{2}{2}\right) = \frac{E_0^2\,2mA^2}{\hbar^2}\left(\frac{4}{4}\right) = \frac{E_0^2}{4E_k}$$

Because $E_0 = \hbar\omega/2$ also $E_0 = 4E_k$

(2) $\partial^2\Psi_0/\partial x^2$ is computed in Problem 6-34(b). Using that quantity,

$$\langle E_k \rangle = -\frac{\hbar^2}{2m}\left(-\frac{m\omega}{2\hbar}\right) = \hbar\omega/4 = 2E_k$$

6-41. (a) For $x>0$, $\hbar^2 k_2^2/2m + V_0 = E = \hbar^2 k_1^2/2m = 2V_0$

So, $k_2 = (2mV_0)^{1/2}/\hbar$. Because $k_1 = (4mV_0)^{1/2}/\hbar$, then $k_2 = k_1/\sqrt{2}$.

(b) $R = (k_1 - k_2)^2/(k_1 + k_2)^2$ (Equation 6-68)

$$= (1 - 1/\sqrt{2})^2/(1 + 1/\sqrt{2})^2 = 0.0294, \ \text{or 2.94\% of the incident particles are reflected.}$$

(Problem 6-41 continued)

(c) $T = 1 - R = 1 - 0.0294 = 0.971$

(d) 97.1% of the particles, or $0.971 \times 10^6 = 9.71 \times 10^5$, continue past the step in the $+x$ direction. Classically, 100% would continue on.

6-45. $A + B = C$ and $k_1 A - k_1 B = k_2 C$ (Equation 6-65 a & b)

Substituting for C, $k_1 A - k_1 B = k_2 (A + B) = k_2 A + k_2 B$ and solving for B,

$B = \dfrac{k_1 - k_2}{k_1 + k_2} A$, which is Equation 6-66. Substituting this value of B into Equation 6-65(a),

$$A + \frac{k_1 - k_2}{k_1 + k_2} A = C = A \left[\frac{k_1 + k_2 + k_1 - k_2}{k_1 + k_2} \right] \quad \text{or} \quad C = \frac{2 k_1}{k_1 + k_2}, \text{ which is Equation 6-67.}$$

6-49. (a) The probability density for the ground state is $P(x) = \psi^2(x) = (2/L)\sin^2 \pi x/L$. The probability of finding the particle in the range $0 < x < L/2$ is :

$$P = \int_0^{L/2} P(x)\,dx = \frac{2}{L}\frac{L}{\pi}\int_0^{\pi/2} \sin^2 u \, du = \frac{2}{\pi}\left(\frac{\pi}{4} - 0 \right) = \frac{1}{2} \quad \text{where } u = \pi x/L$$

(b) $$P = \int_0^{L/3} P(x)\,dx = \frac{2}{L}\frac{L}{\pi}\int_0^{\pi/3} \sin^2 u \, du = \frac{2}{\pi}\left(\frac{\pi}{6} - \frac{\sin 2\pi/3}{4} \right) = \frac{1}{3} - \frac{\sqrt{3}}{4\pi} = 0.195$$

(Note 1/3 is the classical result.)

(c) $$P = \int_0^{3L/4} P(x)\,dx = \frac{2}{L}\frac{L}{\pi}\int_0^{3\pi/4} \sin^2 u \, du = \frac{2}{\pi}\left(\frac{3\pi}{8} - \frac{\sin 3\pi/2}{4} \right) = \frac{3}{4} + \frac{1}{2\pi} = 0.909$$

(Note 3/4 is the classical result.)

6-53.　(a) The requirement is that $\psi^2(x) = \psi^2(-x) = \psi(-x)\psi(-x)$. This can be true only if:

$\psi(-x) = \psi(x)$　or　$\psi(-x) = -\psi(x)$.

(b) Writing the Schrödinger equation in the form $\dfrac{d^2\psi}{dx^2} = -\dfrac{2mE}{\hbar^2}\psi$, the general solutions

of this 2nd order differential equation are:

$\psi(x) = A\sin kx$　and　$\psi(x) = A\cos kx$ where $k = \sqrt{2mE}/\hbar$. Because the boundaries of
the box are at $x = \pm L/2$, both solutions are allowed (unlike the treatment in the text where
one boundary was at $x = 0$). Still, the solutions are all zero at $x = \pm L/2$ provided that an
integral number of half wavelengths fit between $x = -L/2$ and $x = +L/2$. This will occur

for: $\psi_n(x) = (2/L)^{1/2}\cos n\pi x/L$ when $n = 1, 3, 5, \cdots$

And for $\psi_n(x) = (2/L)^{1/2}\sin n\pi x/L$　when $n = 2, 4, 6, \cdots$.

The solutions are alternately even and odd.

(c) The allowed energies are: $E = \hbar^2 k^2/2m = \hbar^2(n\pi L)^2/2m = n^2 h^2/8mL^2$.

6-57.　(a) For $\Psi(x,t) = A\sin(kx - \omega t)$

$\dfrac{d^2\Psi}{dx^2} = -k^2\Psi$　and　$\dfrac{\partial\Psi}{\partial t} = -\omega A\cos(kx - \omega t)$　so the Schrödinger equation becomes:

$-\dfrac{\hbar^2 k^2}{2m}A\sin(kx - \omega t) + V(x)A\sin(kx - \omega t) = -i\hbar\omega\cos(kx - \omega t)$

Because the *sin* and *cos* are not proportional, this Ψ cannot be a solution. Similarly, for
$\Psi(x,t) = A\cos(kx - \omega t)$, there are no solutions.

(b) For $\Psi(x,t) = A[\cos(kx - \omega t) + i\sin(kx - \omega t)] = Ae^{i(kx - \omega t)}$, we have that

$\dfrac{d^2\Psi}{dx^2} = -k^2\Psi$　and　$\dfrac{\partial\Psi}{\partial t} = -i\omega\Psi$. And the Schrödinger equation becomes:

$-\dfrac{\hbar^2 k^2}{2m}\Psi + V(x)\Psi = -\hbar\omega\Psi$ for $\hbar\omega = \hbar^2 k^2/2m + V$.

6-61.   (a) Applying the boundary conditions of continuity to ψ and $d\psi/dx$ at $x = 0$ and $x = a$, where the various wave functions are given by Equations 6-74, results in the two pairs of equations below:

At $x = 0$: $A + B = C + D$ and $ikA - ikB = -\alpha C + \alpha D$

At $x = a$: $Fe^{ika} = Ce^{-\alpha a} + De^{\alpha a}$ and $ikFe^{ika} = -\alpha Ce^{-\alpha a} + \alpha De^{\alpha a}$

Eliminating the coefficients C and D from these four equations, a straightforward but lengthy task, yields:

$*\ \ 4ik\alpha A = \left[(\alpha + ik)^2 e^{-\alpha a} - (\alpha - ik)^2 e^{\alpha a}\right] Fe^{ika}$

The transmission coefficient T is then:

$$T = \frac{|F|^2}{|A|^2} = \left\{ \frac{4ik\alpha}{e^{ika}\left[(\alpha + ik)^2 e^{-\alpha a} - (\alpha - ik)^2 e^{\alpha a}\right]} \right\}^2$$

Recalling that $\sinh\theta = \dfrac{1}{2}\left(e^{\theta} - e^{-\theta}\right)$ and noting that $(\alpha + ik)$ and $(\alpha - ik)$ are complex conjugates, substituting $k = \sqrt{2mE}/\hbar$ and $\alpha = \sqrt{2m(V_0 - E)}/\hbar$, T then can be written as

$$T = \left[1 + \frac{\sinh^2 \alpha a}{4\dfrac{E}{V_0}\left(1 - \dfrac{E}{V_0}\right)} \right]^{-1}$$

(b) If $\alpha a \gg 1$, then the first term in the bracket on the right side of the * equation in part (a) is much smaller than the second and we can write:

$$\frac{F}{A} \approx \frac{4ik\alpha e^{-(\alpha + ik)a}}{(\alpha - ik)^2} \ \ \text{And}\ \ T = \left|\frac{F}{A}\right|^2 \approx \frac{16\alpha^2 k^2 e^{-2\alpha a}}{\left(\alpha^2 + k^2\right)^2}$$

$$\text{Or}\ \ T \approx 16\left(\frac{E}{V_0}\right)\left(1 - \frac{E}{V_0}\right) e^{-2\alpha a}$$

7-1. $E_{n_1 n_2 n_3} = \dfrac{\hbar^2 \pi^2}{2mL^2}\left(n_1^2 + n_2^2 + n_3^2\right)$ (Equation 7-4)

$$E_{311} = \frac{\hbar^2 \pi^2}{2mL^2}\left(3^2 + 1^2 + 1^2\right) = 11E_0 \quad \text{where } E_0 = \frac{\hbar^2 \pi^2}{2mL^2}$$

$$E_{222} = E_0\left(2^2 + 2^2 + 2^2\right) = 12E_0 \quad \text{and} \quad E_{321} = E_0\left(3^2 + 2^2 + 1^2\right) = 14E_0$$

The 1st, 2nd, 3rd, and 5th excited states are degenerate.

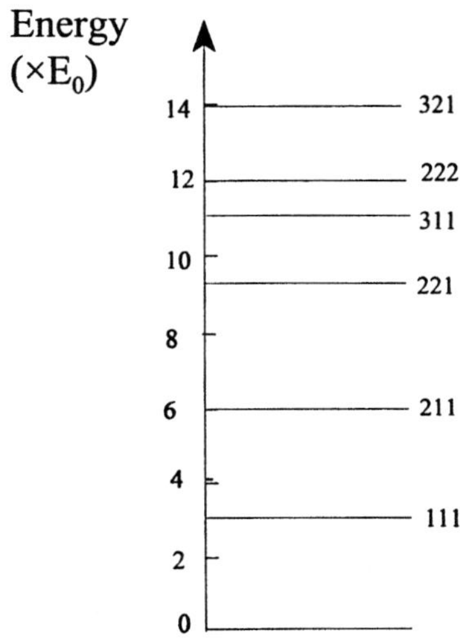

7-5. $E_{n_1 n_2 n_3} = \dfrac{\hbar^2 \pi^2}{2m}\left(\dfrac{n_1^2}{L_1^2} + \dfrac{n_2^2}{\left(2L_1\right)^2} + \dfrac{n_3^2}{\left(4L_1\right)^2}\right) = \dfrac{\hbar^2 \pi^2}{2mL_1^2}\left(n_1^2 + \dfrac{n_2^2}{4} + \dfrac{n_3^2}{16}\right)$ (from Equation 7-5)

$$E_0 = \left(n_1^2 + \frac{n_2^2}{4} + \frac{n_3^2}{16}\right) \quad \text{where } E_0 = \frac{\hbar^2 \pi^2}{2mL_1^2}$$

(Problem 7-5 continued)

(a)

n_1	n_2	n_3	$E \, (\times E_0)$
1	1	1	1.313
1	1	2	1.500
1	1	3	1.813
1	2	1	2.063
1	1	4	2.250
1	2	2	2.250
1	2	3	2.563
1	1	5	2.813
1	2	4	3.000
1	1	6	3.500

(b) 1,1,4 and 1,2,2

7-9. (a) For $n = 3$, $\ell = 0, 1, 2$

(b) For $\ell = 0$, m = 0

$\ell = 1$, m = $-1, 0, +1$

$\ell = 2$, m = $-2, -1, 0 +1, +2$

(c) There are nine different m-states, each with two spin states, for a total of 18 states for $n = 3$.

7-13. $L^2 = L_x^2 + L_y^2 + L_z^2 \;\rightarrow\; L_x^2 + L_y^2 = L^2 - L_z^2 = \ell(\ell+1)\hbar^2 - (m\hbar)^2 = (6 - m^2)\hbar^2$

(a) $\left(L_x^2 + L_y^2\right)_{\min} = (6 - 2^2)\hbar^2 = 2\hbar^2$

(b) $\left(L_x^2 + L_y^2\right)_{\max} = (6 - 0^2)\hbar^2 = 6\hbar^2$

(c) $L_x^2 + L_y^2 = (6 - 1)\hbar^2 = 5\hbar^2$ L_x and L_y cannot be determined separately.

(d) n = 3

7-17. (a) $6f$ state: $n = 6$, $\ell = 3$

  (b) $E_6 = -13.6\,eV/n^2 = -13.6\,eV/6^2 = -0.38\,eV$

  (c) $L = \sqrt{\ell(\ell+1)}\,\hbar = \sqrt{3(3+1)}\,\hbar = \sqrt{12}\,\hbar = 3.65 \times 10^{-34}\,J\cdot s$

  (d) $L_z = m\hbar$ $L_z = -3\hbar,\ -2\hbar,\ -1\hbar,\ 0,\ 1\hbar,\ 2\hbar,\ 3\hbar$

7-21. $P(r) = Cr^2 e^{-2Zr/a_o}$ For P(r) to be a maximum,

$$\frac{dP}{dr} = C\left[r^2\left(-\frac{2Z}{a_o}\right)e^{-2Zr/a_o} + 2re^{-2Zr/a_o}\right] = 0 \qquad \rightarrow \qquad C\cdot\frac{2Zr}{a_o}\left(\frac{a_o}{Z} - r\right)e^{-2Zr/a_o} = 0$$

This condition is satisfied when $r = 0$ or $r = a_o/Z$. For r = 0, P(r) = 0 so the maximum P(r) occurs for $r = a_o/Z$.

7-25. $\psi_{200} = \dfrac{1}{\sqrt{32\pi}}\left(\dfrac{1}{a_0}\right)^{3/2}\left(2 - \dfrac{r}{2a_0}\right)e^{-r/2a_0}$ (Z = 1 for hydrogen)

  (a) At r = a_0,

$$\psi_{200} = \frac{1}{\sqrt{32\pi}}\left(\frac{1}{a_0^3}\right)(2-1)e^{-1/2} = \frac{0.606}{\sqrt{32\pi}}\left(\frac{1}{a_0}\right)^{3/2}$$

  (b) At r = a_0,

$$\left|\psi_{200}\right|^2 = \frac{1}{\sqrt{32\pi}}\left(\frac{1}{a_0^3}\right)e^{-1} = \frac{0.368}{32\pi}\frac{1}{a_0^3}$$

  (c) At r = a_0,

$$P(r) = \left|\psi_{200}\right|^2(4\pi r^2) = \frac{4\pi}{32\pi}\frac{0.368\,a_0^2}{a_0^3} = \frac{0.368}{8a_0}$$

7-29. (a) Every increment of charge follows a circular path of radius R and encloses an area πR^2, so the magnetic moment is the total current times this area. The entire charge Q rotates with frequency $f = \omega/2\pi$, so the current is

(Problem 7-29 continued)

$$i = Qf = q\omega/2\pi$$

$$\mu = iA = (Q\omega/2\pi)(\pi R^2) = Q\omega R^2/2$$

$$L = I\omega = \frac{1}{2}MR^2\omega$$

$$g = \frac{2M\mu}{QL} = \frac{2MQ\omega R^2/2}{QMR^2\omega/2} = 2$$

(b) The entire charge is on the equatorial ring, which rotates with frequency $f = \omega/2\pi$.

$$i = Qf = Q\omega/2\pi$$

$$\mu = iA = (Q\omega/2\pi)(\pi R^2) = Q\omega R^2/2$$

$$g = \frac{2M\mu}{QL} = \frac{2MQ\omega R^2/2}{QMR^2\omega/5} = 5/2 = 2.5$$

7-33. (a) There should be four lines corresponding to the four m_j values $-3/2, -1/2, +1/2, +3/2$.

(b) There should be three lines corresponding to the three m_ℓ values $-1, 0, +1$.

7-37. $j = \ell \pm 1/2$. $\ell = j \pm 1/2 = 3/2 \pm 1/2 = 1 \ or \ 2$

7-41. $\psi_{12} = \psi(x_1, x_2) = C\sin\dfrac{\pi x_1}{L}\sin\dfrac{2\pi x_2}{L}$ Substituting into Equation 7-57 with $V = 0$,

$$-\frac{\hbar^2}{2m}\left(\frac{\partial^2\psi_{12}}{\partial x_1^2} + \frac{\partial^2\psi_{12}}{\partial x_2^2}\right) = \left(\frac{\hbar^2}{2m}\right)(1+4)\left(\frac{\pi^2}{L^2}\right)\psi_{12} = E\psi_{12}$$

Obviously, ψ_{12} is a solution if $E = \dfrac{5\hbar^2\pi^2}{2mL^2}$

7-45. (a) Chlorine: $Z = 17$; $1s^2 2s^2 2p^6 3s^2 3p^5$

(b) Calcium: $Z = 20$; $1s^2 2s^2 2p^6 3s^2 3p^6 4s^2$

(c) Germanium : $Z = 32$; $1s^2 2s^2 2p^6 3s^2 3p^6 3d^{10} 4s^2 4p^2$

7-49. (a) Fourteen electrons, so $Z = 14$. Element is silicon.

(b) Twenty electrons. So $Z = 20$. Element is calcium.

7-53. Similar to H: Li, Rb, Ag, and Fr. Similar to He: Ca, Ti, Cd, Ba, Hg, and Ra.

7-57. $\Delta j = \pm 1, 0$ (no $j = 0 \rightarrow j = 0$) (Equation 7-66)

The four states are $^2P_{3/2}$, $^2P_{1/2}$, $^2D_{5/2}$, $^2D_{3/2}$.

Transition	$\Delta \ell$	Δj	Comment
$D_{5/2} \rightarrow P_{3/2}$	-1	-1	allowed
$D_{5/2} \rightarrow P_{1/2}$	-1	-2	j - forbidden
$D_{3/2} \rightarrow P_{3/2}$	-1	0	allowed
$D_{3/2} \rightarrow P_{1/2}$	-1	-1	allowed

7-61. (a) $\Delta E = \dfrac{e \hbar}{2m} B = (5.79 \times 10^{-4}\ eV/T)(0.005\ T) = 2.90 \times 10^{-5}\ eV$

(b) $|\Delta \lambda| = \dfrac{\lambda^2}{hc} \Delta E = \dfrac{(579.07\ nm)^2 (2.90 \times 10^{-5}\ eV)}{1240\ eV \cdot nm} = 7.83 \times 10^{-3}\ nm$

(c) The smallest measurable wavelength change is larger than this by the ratio
0.01 nm / 0.00783 nm = 1.28. The magnetic field would need to be increased by this
same factor because $B \propto \Delta E \propto \Delta \lambda$. The necessary field would be 0.0638 T.

7-65. (a) $|F_z| = m_s g_L \mu_B (dB/dz)$ (From Equation 7-51)

From Newton's 2nd law,

$$|F_z| = m_H a_z = m_s g_L \mu_B (dB/dz)$$

$$a_z = m_s g_L (dB/dz)/m_H$$

$$= (1/2)(1)(9.27 \times 10^{-24}\,J/T)(600\,T/m)/(1.67 \times 10^{-27}\,kg)$$

$$= 1.67 \times 10^6\,m/s^2$$

(b) At 14.5 km/s $= v = 1.45 \times 10^4\,m/s$, the atom takes $t_1 = 0.75\,m/(1.45 \times 10^4\,m/s) = 5.2 \times 10^{-5}\,s$

to traverse the magnet. In that time, its z deflection will be:

$$z_1 = (1/2)(a_z)t_1^2 = (1/2)(1.67 \times 10^6\,m/s^2)(5.2 \times 10^{-5}\,s)^2 = 2.26 \times 10^{-3}\,m = 2.26\,m$$

Its v_z velocity component as it leaves the magnet is $v_z = a_z t_1$ and its additional z

deflection before reaching the detector 1.25 m away will be:

$$z_2 = v_z t_2 = (a_z t_1)(1.25\,m/[1.45 \times 10^4\,m/s])$$

$$= (1.67 \times 10^6\,m/s^2)(5.2 \times 10^{-5}\,s)(1.25)/(1.45 \times 10^4\,m/s)$$

$$= 7.49 \times 10^{-3}\,m = 7.49\,mm$$

Each line will be deflected $z_1 + z_2 = 9.75\,mm$ from the central position and, thus,

separated by a total of 19.5 mm = 1.95 cm.

7-69. (a) $g = 1 + \dfrac{j(j+1) + s(s+1) - \ell(\ell+1)}{2j(j+1)}$ (Equation 7-73)

For $^2P_{1/2}$: $j = 1/2$, $\ell = 1$, and $s = \frac{1}{2}$

$$g = 1 + \frac{1/2(1/2+1) + 1/2(1/2+1) - 1(1+1)}{2 \cdot 1/2(1/2+1)} = 1 + \frac{3/4 + 3/4 - 2}{3/2} = 2/3$$

For $^2S_{1/2}$: $j = 1/2$, $\ell = 0$, and $s = \frac{1}{2}$

$$g = 1 + \frac{1/2(1/2+1) + 1/2(1/2+1) - 0}{2 \cdot 1/2(1/2+1)} = 1 + \frac{3/4 + 3/4}{3/2} = 2$$

(Problem 7-69 continued)

The $^2P_{1/2}$ levels shift by:

$$\Delta E = g m_j \mu_B B = \frac{2}{3}\left(\pm\frac{1}{2}\right)\mu_B B = \pm\frac{1}{3}\mu_B B \quad \text{(Equation 7-72)}$$

The $^2S_{1/2}$ levels shift by:

$$\Delta E = g m_j \mu_B B = 2\left(\pm\frac{1}{2}\right)\mu_B B = \pm\,\mu_B B$$

To find the transition energies, tabulate the several possible transitions and the corresponding energy values (Let E_p and E_s be the $B = 0$ unsplit energies of the two states.):

Transition	Energy

$P_{1/2,\,1/2} \rightarrow S_{1/2,\,1/2}$

$$\left(E_p + \frac{1}{3}\mu_B B\right) - (E_s + \mu_B B) = (E_p - E_s) - \frac{2}{3}\mu_B B$$

$P_{1/2,\,-1/2} \rightarrow S_{1/2,\,1/2}$

$$\left(E_p - \frac{1}{3}\mu_B B\right) - (E_s + \mu_B B) = (E_p - E_s) - \frac{4}{3}\mu_B B$$

$P_{1/2,\,1/2} \rightarrow S_{1/2,\,-1/2}$

$$\left(E_p + \frac{1}{3}\mu_B B\right) - (E_s - \mu_B B) = (E_p - E_s) + \frac{4}{3}\mu_B B$$

Transition	Energy

$P_{1/2,\,-1/2} \rightarrow S_{1/2,\,-1/2}$

$$\left(E_p - \frac{1}{3}\mu_B B\right) - (E_s - \mu_B B) = (E_p - E_s) + \frac{2}{3}\mu_B B$$

Thus, there are four different photon energies emitted. The energy or frequency spectrum would appear as below (normal Zeeman spectrum shown for comparison).

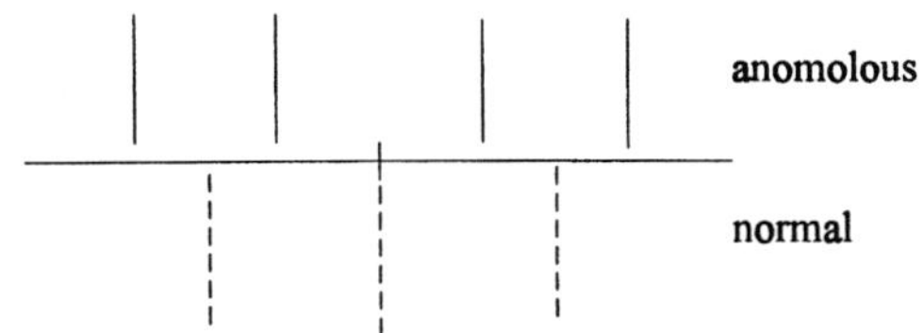

(b) For $^2P_{3/2}$: $j = 3/2$, $\ell = 1$, and $s = \frac{1}{2}$

(Problem 7-69 continued)

$$g = 1 + \frac{3/2(3/2+1) + 1/2(1/2+1) - 1(1+1)}{2 \cdot 3/2(3/2+1)} = 1 + \frac{15/4 + 3/4 - 2}{30/4} = 4/3$$

These levels shift by:

$$\Delta E = g m_j \mu_B B = \frac{4}{3}\left(\pm\frac{1}{2}\right)\mu_B B = \pm\frac{2}{3}\mu_B B \qquad \Delta E = \frac{4}{3}\left(\pm\frac{3}{2}\right)\mu_B B = \pm 2\mu_B B$$

Tabulating the transitions as before:

<u>Transition</u>	<u>Energy</u>
$P_{3/2,\,3/2} \rightarrow S_{1/2,\,1/2}$	$\left(E_p + 2\mu_B B\right) - \left(E_s + \mu_B B\right) = \left(E_p - E_s\right) + \mu_B B$
$P_{3/2,\,3/2} \rightarrow S_{1/2,\,-1/2}$	forbidden, $\Delta m_j = 2$
$P_{3/2,\,1/2} \rightarrow S_{1/2,\,1/2}$	$\left(E_p - \frac{2}{3}\mu_B B\right) - \left(E_s + \mu_B B\right) = \left(E_p - E_s\right) - \frac{1}{3}\mu_B B$
$P_{3/2,\,1/2} \rightarrow S_{1/2,\,-1/2}$	$\left(E_p + \frac{2}{3}\mu_B B\right) - \left(E_s - \mu_B B\right) = \left(E_p - E_s\right) + \frac{5}{3}\mu_B B$
$P_{3/2,\,-1/2} \rightarrow S_{1/2,\,1/2}$	$\left(E_p - \frac{2}{3}\mu_B B\right) - \left(E_s + \mu_B B\right) = \left(E_p - E_s\right) - \frac{5}{3}\mu_B B$
$P_{3/2,\,-1/2} \rightarrow S_{1/2,\,-1/2}$	$\left(E_p - \frac{2}{3}\mu_B B\right) - \left(E_s - \mu_B B\right) = \left(E_p - E_s\right) + \frac{1}{3}\mu_B B$
$P_{3/2,\,-3/2} \rightarrow S_{1/2,\,1/2}$	forbidden $\Delta m_j = 2$
$P_{3/2,\,-3/2} \rightarrow S_{1/2,\,-1/2}$	$\left(E_p - 2\mu_B B\right) - \left(E_s - \mu_B B\right) = \left(E_p - E_s\right) - \mu_B B$

There are six different photon energies emitted (two transitions are forbidden); their spectrum looks as below:

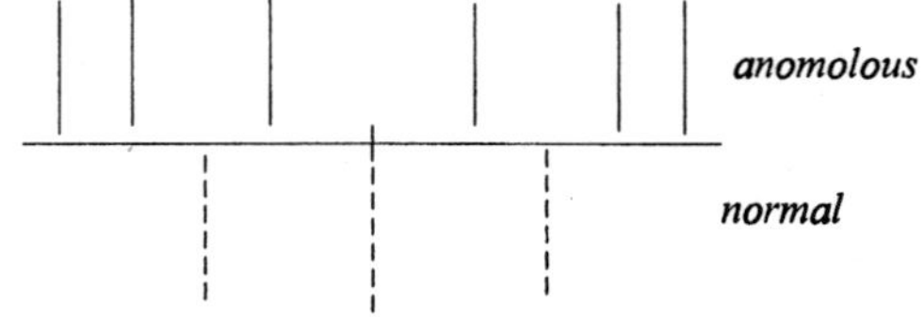

7-73. (a) $\mathbf{J} = \mathbf{L} + \mathbf{S}$ $\boldsymbol{u} = -\mu_B(\mathbf{L} + 2\mathbf{S})/\hbar$ (Equation 7-71)

$$u_J = \frac{\boldsymbol{u}\cdot\boldsymbol{J}}{J} = \frac{\left[-\mu_B(\mathbf{L}+2\mathbf{S})/\hbar\right]\cdot\left[\mathbf{L}+\mathbf{S}\right]}{J}$$

$$\equiv -\frac{\mu_B}{\hbar J}\left(\mathbf{L}\cdot\mathbf{L} + 2\mathbf{S}\cdot\mathbf{S} + 3\mathbf{S}\cdot\mathbf{L}\right)$$

$$= -\frac{\mu_B}{\hbar J}\left(L^2 + 2S^2 + 3\mathbf{S}\cdot\mathbf{L}\right)$$

(b) $J^2 = \boldsymbol{J}\cdot\boldsymbol{J} = (\mathbf{L}+\mathbf{S})\cdot(\mathbf{L}+\mathbf{S}) = \mathbf{L}\cdot\mathbf{L} + \mathbf{S}\cdot\mathbf{S} + 2\mathbf{S}\cdot\mathbf{L}$ $\therefore\ \mathbf{S}\cdot\mathbf{L} = \frac{1}{2}\left(J^2 - L^2 - S^2\right)$

(c) $\mu_J = -\frac{\mu_B}{\hbar J}\left[L^2 + 2S^2 + \frac{3}{2}\left(J^2 - L^2 - S^2\right)\right] = -\frac{\mu_B}{2\hbar J}\left(3J^2 + S^2 - L^2\right)$

(d) $\mu_Z = \mu_J\dfrac{J_Z}{J} = -\dfrac{\mu_B}{2\hbar J}\left(3J^2 + S^2 - L^2\right)\dfrac{J_Z}{J} = -\dfrac{\mu_B}{2\hbar J}\left(3J^2 + S^2 - L^2\right)$

$$= -\mu_B\left(1 + \frac{J^2 + S^2 - L^2}{2J^2}\right)\frac{J_Z}{\hbar}$$

(e) $\Delta E = -\mu_Z B$ (Equation 7-69)

$$= +\mu_B B\left[1 + \frac{j(j+1) + s(s+1) - \ell(\ell+1)}{2j(j+1)}\right]m_j$$

$$= g\,m_j\,\mu_B B \quad \text{(Equation 7-72)}$$

$$\text{where}\quad g = \left[1 + \frac{j(j+1) + s(s+1) - \ell(\ell+1)}{2j(j+1)}\right] \quad \text{(Equation 7-73)}$$

Chapter 8 – Statistical Physics

8-1. (a) $v_{rms} = \sqrt{\dfrac{3RT}{M}} = \left[\dfrac{3(8.31\ J/mole\cdot K)(300K)}{2(1.0079\times10^{-3}\ kg/mole)}\right]^{1/2} = 1930\ m/s$

   (b) $T = \dfrac{Mv_{rms}^2}{3R} = \dfrac{2(1.0079\times10^{-3}\ kg/mole)(11.2\times10^3\ m/s)^2}{3(8.31\ J/mole\cdot K)} = 1.01\times10^4\ K$

8-5. (a) $E_K = n\cdot\dfrac{3}{2}RT = (1\ mole)\dfrac{3}{2}(8.31\ J/mole\cdot K)(273) = 3400\ J$

   (b) One mole of any gas has the same translational kinetic energy at the same temperature.

8-9. $m(v)\,dv = 4\pi N\left(\dfrac{m}{2\pi kT}\right)^{3/2} v^2 e^{-mv^2/2kT}\,dv$ (Equation 8-28)

   $\dfrac{du}{dv} = A\left[v^2\left(-\dfrac{2vm}{2kT}\right) + 2v\right]e^{-mv^2/2kT}$ The v for which $dn/dv = 0$ is v_m.

   $A\left[-\dfrac{2mv^3}{2kT} + 2v\right]e^{-mv^2/2kT} = 0$

   Because A = constant and the exponential term is only zero for $v = \infty$, only the

   quantity in [] can be zero, so $-\dfrac{2mv^3}{2kT} + 2v = 0$

   or $v^2 = \dfrac{2kT}{m} \rightarrow v_m = \sqrt{\dfrac{2kT}{m}}$ (Equation 8-29)

8-13. There are two degrees of freedom; therefore, $C_v = 2(R/2) = R$, $C_p = R + R = 2R$, and

$\gamma = 2R/R = 2$.

8-17. $n(E) = Ag(E)e^{-E/kT}$ (Equation 8-14)

The degeneracies of the lowest four hydrogen states are:

$n = 1$, $g(E_1) = 2$; $n = 2$, $g(E_2) = 8$; $n = 3$, $g(E_3) = 18$; $n = 4$, $g(E_4) = 32$

For the Sun $kT = 0.500\,eV$.

$$\frac{n(E_2)}{n(E_1)} = \frac{Ag_2 e^{-E_2/kT}}{Ag_1 e^{-E_1/kT}} = \frac{8}{2}e^{-(E_2-E_1)/kT} = 4e^{-(10.2)/(0.500)} = 4e^{-20.4} = 5.5 \times 10^{-9}$$

$$\frac{n(E_3)}{n(E_1)} = \frac{18}{2}e^{-(12.1)/(0.500)} = 9e^{-24.2} = 2.8 \times 10^{-10}$$

$$\frac{n(E_4)}{n(E_1)} = \frac{32}{2}e^{-(12.8)/(0.500)} = 16e^{-25.5} = 1.3 \times 10^{-10}$$

8-21. (a) $f_{BE} = \dfrac{1}{e^{\alpha}\,e^{E/kT} - 1}$ For $\alpha = 0$ and $f_{BE} = 1$, at $T = 5800$ K

$$\frac{1}{e^{E/5800k} - 1} = 1 \;\rightarrow\; e^{E/5800k} = 2$$

$$E/(5800K)(8.62 \times 10^{-5}\,eV/K) = \ln 2$$

$$E = 0.347 \text{ eV}$$

(b) For E = 0.35 V, $\alpha = 0$, and $f_{BE} = 0.5$,

$$\frac{1}{e^{0.35/kT} - 1} = 0.5 \;\rightarrow\; e^{0.35/kT} = 3$$

$$0.35\,eV / (8.67 \times 10^{-5\,eV/K})T = \ln 3$$

$$T = \frac{0.35\,eV}{\ln 3\,(8.67 \times 10^{-5}\,eV/K)} = 3660\,K$$

8-25. $\quad T_C = \dfrac{h^2}{2mk}\left[\dfrac{N}{2\pi(2.315)V}\right]^{2/3}$ (Equation 8-72)

The density of liquid Ne is 1.207 g/cm^3, so

$$\frac{N}{V} = \frac{\left(1.207\,g/cm^3\right)\left(6.022\times10^{23}\,molecules/mol\right)\left(10^6\,cm^2/m^3\right)}{20.18\,g/mol} = 3.601\times10^{28}\ /m^3$$

$$T_C = \frac{\left(6.626\times10^{-34}\,J\cdot s\right)^2}{2\left(20\,u\times1.66\times10^{-27}\,kg/u\right)\left(1.381\times10^{-23}\,J/K\right)}\left[\frac{3.601\times10^{28}\,m^3}{2\pi(2.315)}\right]^{2/3} = 0.895\ K$$

Thus, T_C at which ^{20}Ne would become a superfluid is much lower than its freezing temperature of 24.5 K.

8-29. $\quad C_V = 3R\left(\dfrac{hf}{kT}\right)^2\dfrac{e^{hf/kT}}{\left(e^{hf/kT}-1\right)^2}$ (Equation 8-86)

Writing $hf/kT = Af$ where $A = h/kT = 2.40\times10^{-13}$ when $T = 200\,K$,

$$C_V = 3R(Af)^2\frac{e^{Af}}{\left(e^{Af}\right)^2 - 2e^{Af} + 1} = eR(Af)^2\frac{1}{e^{Af} - 2 + \left(1/e^{Af}\right)}$$

Because Af is "large", $1/e^{Af} \approx 0$ and e^{Af} dominates $(Af)^2$, so

$$C_V \approx 3R/e^{Af} \rightarrow e^{Af} \approx 3R/C_V \rightarrow f \approx \ln(3R/C_V)(1/A)$$

For Al, $C_V(200K) = 20.1\,J/K\cdot mol$ (From Figure 8-14)

$$f = \ln\left(\frac{3(8.31)}{20.1}\right)\left(1/2.40\times10^{-13}\right) = 8.97\times10^{11}\ Hz$$

For Is, $C_V(200K) = 13.8\,J/K\cdot mol$ (From Figure 8-14)

$$f = \ln\left(\frac{3(8.31)}{13.8}\right)\left(1/2.40\times10^{-13}\right) = 2.46\times10^{12}\ Hz$$

8-33. Approximating the nuclear potential with an infinite square well and ignoring the Coulomb repulsion of the protons, the energy levels for both protons and neutrons are given by $E_n = (n^2 h^2)/(8mL^2)$ and six levels will be occupied in ^{22}Ne, five levels with 10 protons and six levels with 12 neutrons.

$$E_F(\text{protons}) = \frac{(5)^2(1240\ MeV \cdot fm)^2}{8(1.0078\,u \times 931.5\ MeV/u)(3.15\,fm)^2} = 516\ MeV$$

$$E_F(\text{neutrons}) = \frac{(6)^2(1240\ MeV \cdot fm)^2}{8(1.0087\,u \times 931.5\ MeV/u)(3.15\,fm)^2} = 742\ MeV$$

$$\langle E \rangle (\text{protons}) = (3/5)E_F = 310\ MeV$$

$$\langle E \rangle (\text{neutrons}) = (3/5)E_F = 445\ MeV$$

As we will discover in Chapter 11, these estimates are nearly an order of magnitude too large. The number of particles is not a large sample.

8-37. For a one-dimensional well approximation, $E_n = (n^2 h^2)/(8mL^2)$. At the Fermi level E_F, $n = N/2$, where N = number of electrons.

$$E_F = \frac{(N/2)^2 h^2}{8mL^2} = \frac{h^2}{32m}\left(\frac{N}{L}\right)^2 \qquad \text{where } N/L = \text{number of electrons/unit length,}$$

i.e., the density of electrons. Assuming 1 free electron/Au atom,

$$\frac{N}{L} = \left[\frac{N_A \rho}{M}\right]^{1/3} = \left[\frac{6.02 \times 10^{23}\ electrons/mol\ (19.32\,g/cm^3)(10^2\,cm/m)^3}{197\ g/mol}\right]^{1/3} = 3.81 \times 10^9\ m^{-1}$$

$$E_F = \frac{(6.63 \times 10^{-34}\,J \cdot s)^2 (3.81 \times 10^9\,m^{-1})^2}{8(9.11 \times 10^{-31}\,kg)(1.60 \times 10^{-19}\,J/eV)} = 647\ eV$$

This is the energy of an electron in the Fermi level above the bottom of the well. Adding the work function to such an electron just removes it from the metal, so the well is $5.47\,eV + 4.8\,eV = 10.3\,eV$ deep.

8-41. (a) $f(u)\,du = Ce^{-E/kT}\,du = Ce^{-Au^2/kT}\,du$ (from Equation 8-17)

$$1 = \int_{-\infty}^{+\infty} f(u)\,du = \int_{-\infty}^{+\infty} Ce^{-Au^2/kT}\,du = 2C\int_0^{\infty} e^{-Au^2/kT}\,du$$

$$= 2CI_0 = 2C\sqrt{\pi}\,\lambda^{-1/2}/2 \quad \text{where } \lambda = A/kT$$

$$= C\sqrt{\pi}\,\sqrt{kT/A} \quad \rightarrow \quad C = \sqrt{A/\pi kT}$$

(b) $\langle E \rangle = \langle Au^2 \rangle = \displaystyle\int_{-\infty}^{+\infty} Au^2 f(u)\,du = \int_{-\infty}^{+\infty} Au^2 \sqrt{A/\pi kT}\; e^{-Au^2/kT}\,du$

$$= A\sqrt{A/\pi kT}\,(2I_2) = A\sqrt{A/\pi kT}\; 2\cdot\left(\sqrt{\pi}/4\right)\lambda^{-3/2} \quad \text{where } \lambda = A/kT$$

$$= \frac{1}{2}A\sqrt{A/kT}\,(kT/A)^{3/2} = \frac{1}{2}kT$$

8-45. (a) $N = \displaystyle\sum_i n_i = f_0(E_0) + f_1(E_1)$ (with $g_0 = g_1 = 1$)

$$= Ce^0 + Ce^{-\epsilon/kT} = C\left(1 + e^{-\epsilon/kT}\right)$$

So, $C = \dfrac{N}{1 + e^{-\epsilon/kT}}$

(b) $\langle E \rangle = \dfrac{0\cdot n_0 + \epsilon n_1}{N} = \dfrac{\epsilon C e^{-\epsilon/kT}}{N} = \dfrac{N\epsilon e^{-\epsilon/kT}}{\left(1 + e^{-\epsilon/kT}\right)N} = \dfrac{\epsilon e^{-\epsilon/kT}}{1 + e^{-\epsilon/kT}}$

As $T \rightarrow 0$, $e^{-\epsilon/kT} = 1/e^{-\epsilon/kT} \rightarrow 0$, so $\langle E \rangle \rightarrow 0$

As $T \rightarrow \infty$, $e^{-\epsilon/kT} = 1/e^{-\epsilon/kT} \rightarrow 1$, so $\langle E \rangle \rightarrow \epsilon/2$

$$C_V = \frac{dE}{dT} = \frac{d(N\langle E \rangle)}{dT} = \frac{d}{dT}\left(\frac{N\epsilon e^{-\epsilon/kT}}{1 + e^{-\epsilon/kT}}\right)$$

(c)

$$= \frac{N\epsilon^2}{kT^2}\left[\frac{-\left(e^{-\epsilon/kT}\right)^2}{\left(1 + e^{-\epsilon/kT}\right)^2} + \frac{e^{-\epsilon/kT}}{\left(1 + e^{-\epsilon/kT}\right)}\right]$$

$$= Nk\left(\frac{\epsilon}{kT}\right)^2 \frac{e^{-\epsilon/kT}}{\left(1 + e^{-\epsilon/kT}\right)^2}$$

(Problem 8-45 continued)

(d)

T ($\times \epsilon/k$)	0.1	0.25	0.5	1.0	2.0	3.0
C_v ($\times Nk$)	0.005	0.28	0.42	0.20	0.06	0.03

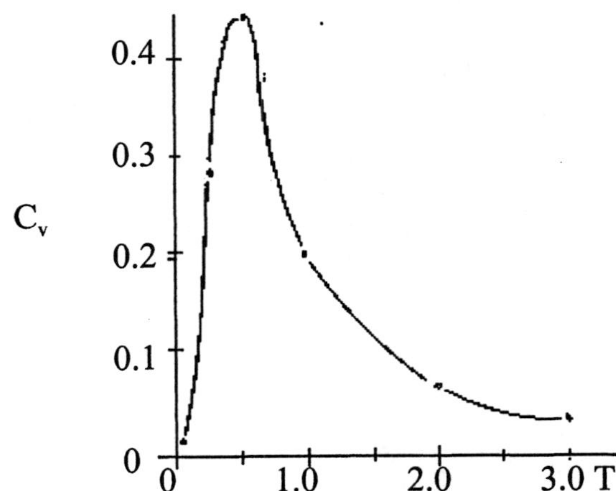

Chapter 9 – Molecular Structure and Spectra

9-1.   (a) $1\dfrac{eV}{molecule} = \left(1\dfrac{eV}{molecule}\right)\left(\dfrac{1.602\times10^{-19}J}{eV}\right)\left(\dfrac{6.022\times10^{23}\,molecules}{mole}\right)$

$$= \left(96472\,\dfrac{J}{mole}\right)\left(\dfrac{1\,cal}{4.184\,J}\right) = 23057\,\dfrac{cal}{mole} = 23.06\,\dfrac{kcal}{mole}$$

(b) $E_d = \left(4.27\,\dfrac{eV}{molecule}\right)\left(\dfrac{23.06\,kcal/mole}{1\,eV/molecule}\right) = 98.5\,kcal/mole$

(c) $E_d = \left(106\,\dfrac{kJ}{mole}\right)\left(\dfrac{1\,eV/molecule}{96.47\,kJ/mole}\right) = 1.08\,eV/molecule$

9-5.   (a) Total potential energy: $U(r) = -\dfrac{ke^2}{r} + E_{ex} + E_{ion}$ (Equation 9-1)

attractive part of $U(r_0) = -\dfrac{ke^2}{r_0} = -\dfrac{1.44\,eV\cdot nm}{0.267\,nm} = -5.39\,eV$

(b) The net ionization energy is:

$E_{ion} = (ionization\ energy\ of\ Rb) - (electron\ affinity\ of\ Cl)$

$= 4.18\,eV - 3.62\,eV = 0.56\,eV$

Neglecting the exclusion principle repulsion energy E_{ex},

$dissociation\ energy\ = -U(r_0) = 5.39\,eV - 0.56\,eV = 4.83\,eV$

(c) Including exclusion principle repulsion,

$dissociation\ energy\ = 4.37\,eV = -U(r_0) = 5.39\,eV - 0.56\,eV - E_{ex}$

$E_{ex} = 5.39\,eV - 4.37\,eV - 0.56\,eV = 0.46\,eV$

9-9. For *KBr*: $U_C = \dfrac{1.440\,eV\cdot nm}{0.282\,nm} + (4.34\,eV - 3.36\,eV) = -4.13\,eV$

$$E_d = 3.94\,eV = \left|U_c + E_{ex}\right| = \left|-4.13\,eV + E_{ex}\right|$$

$$E_{ex} = 0.19\,eV$$

For *RbCl*: $U_C = \dfrac{1.440\,eV\cdot nm}{0.279\,nm} + (4.18\,eV - 3.62\,eV) = -4.60\,eV$

$$E_d = 4.37\,eV = \left|U_c + E_{ex}\right| = \left|-4.60\,eV + E_{ex}\right|$$

$$E_{ex} = 0.23\,eV$$

9-13. $p_{ionic} = er_0 = (1.60 \times 10^{-19}\,C)(0.2345 \times 10^{-9}\,m)$ (Equation 9-3)

$$= 3.757 \times 10^{-29}\,C\cdot m,\ \text{if purely ionic.}$$

The measured value should be:

$$p_{ionic}(\text{measured}) = 0.70\,p_{ionic} = 0.70\,(3.757 \times 10^{-29}\,C\cdot m) = 2.630 \times 10^{-29}\,C\cdot m$$

9-17. $U = \alpha k^2 p_1^2 / r^2$ (Equation 9-10)

(a) Kinetic energy of $N_2 = 0.026\,eV$, so when $|U| = 0.026\,eV$ the bond will be broken.

$$0.026\,eV = \frac{(1.1 \times 10^{-37}\,m\cdot C^2/N)(9 \times 10^9\,N\cdot m^2/C^2)^2(6.46 \times 10^{-30}\,C\cdot m)^2}{r^6}$$

$$r^6 = \frac{(1.1 \times 10^{-37}\,m\cdot C^2/N)(9 \times 10^9\,N\cdot m^2/C^2)^2(6.46 \times 10^{-30}\,C\cdot m)^2}{0.026\,eV(1.60 \times 10^{-19}\,J/eV)}$$

$$= 8.94 \times 10^{-56}\,m^6$$

$$r = 6.7 \times 10^{-10}\,m = 0.67\,nm$$

(b) $U \approx -\dfrac{ke^2}{r} \ \rightarrow\ |U| = 0.026\,eV = \dfrac{1.440\,eV\cdot nm}{r} \ \rightarrow\ r \approx 55\,nm$

(c) $H_2O\text{-}Ne$ bonds in the atmosphere would be very unlikely. The individual molecules will, on the average, be about 4 nm apart, but if a $H_2O\text{-}Ne$ molecule should form, its $U \approx 0.003\,eV$ at $r = 0.95\,nm$, a typical (large) separation. Thus, a N_2 molecule with the average kinetic energy could easily dissociate the $H_2O\text{-}Ne$ bond.

9-21. $E_V = \dfrac{\hbar^2}{2I}$ (Equation 9-14) *where* $I = \dfrac{1}{2}mr_0^2$ for a symmetric molecule.

$$E_V = \frac{\hbar^2}{mr_0^2} = \frac{(\hbar c)^2}{mc^2 r_0^2)} = \frac{(197.3\ eV\cdot nm)^2}{(16\,uc^2)(931.5\times 10^6\ eV/uc^2)(0.121\,nm)^2} = 1.78\times 10^{-4}\ eV$$

9-25. (a) $\mu = \dfrac{m_1 m_2}{m_1 + m_2} = \dfrac{(39.1\,u)(35.45\,u)}{39.1\,u + 35.45\,u} = 18.6\,u$

(b) $E_V = \dfrac{\hbar^2}{2I}$ (Equation 9-14) $I = \mu r_0^2$

$$E_V = \frac{\hbar^2}{2\mu r_0^2} = \frac{(\hbar c)^2}{2\mu c^2 r_0^2} \quad\rightarrow\quad r_0^2 = \frac{(\hbar c)^2}{2\mu c^2 E_V}$$

$$\therefore\quad r_o = \frac{\hbar c}{(2\mu c^2 E_V)^{1/2}} = \frac{197.3\ eV\cdot nm}{\left[2(10.6\,uc^2)(931.5\times 10^6\,eV/uc^2)(1.43\times 10^{-5}\,eV)\right]^{1/2}}$$

$$= 0.280\ nm$$

9-29. $E_{0r} = \dfrac{\hbar^2}{2I}$ where $I = \mu r_0^2$ (Equation 9-14)

For $K^{35}Cl$: $\mu = \dfrac{(39.102\,u)(34.969\,u)}{39.102\,u + 34.969\,u} = 18.46\ u$

For $K^{37}Cl$: $\mu = \dfrac{(39.102\,u)(34.966\,u)}{39.102\,u + 34.966\,u} = 19.00\ u$

$r_0 = 0.267\,nm$ for KCl.

$$E_{0r}(K^{35}Cl) = \frac{\left(1.055\times 10^{-34}\,J\cdot s\right)^2}{2(18.46\,u)(1.66\times 10^{-27}\,kg/u)(0.267\times 10^{-9}\,m)^2}$$

$$= 2.55\times 10^{-24}\,J = 1.59\times 10^{-5}\ eV$$

$$E_{0r}(K^{37}Cl) = \frac{\left(1.055\times 10^{-34}\,J\cdot s\right)^2}{2(19.00\,u)(1.66\times 10^{-27}\,kg/u)(0.267\times 10^{-9}\,m)^2}$$

(Problem 9-29 continued)

$$= 2.48 \times 10^{-24} J = 1.55 \times 10^{-5} eV$$

$$\Delta E_{or} = 0.04 \times 10^{-5} eV$$

9-33. $\dfrac{n(E_1)}{n(E_0)} = \dfrac{e^{-E_1/kT}}{e^{-E_0/kT}}$ i.e., the ratio of the Boltzmann factors.

For O_2 : $f = 4.74 \times 10^{13} Hz$ and

$$E_0 = hf/2 = (4.14 \times 10^{-15} eV \cdot s)(4.74 \times 10^{13} Hz)/2 = 0.0981 eV$$

$$E_1 = 3 hf/2 = 0.294 eV$$

At 273 K, $kT = (8.62 \times 10^{-5} eV/K)(273 K) = 0.0235 eV$

$$\frac{n(E_1)}{n(E_0)} = \frac{e^{-0.294/0.0235}}{e^{-0.0981/0.0235}} = \frac{e^{-12.5}}{e^{-4.17}} = 2.4 \times 10^{-4}$$

Thus, about 2 of every 10,000 molecules are in the E_1 state.

Similarly, at 77K, $\dfrac{n(E_1)}{n(E_0)} = 1.4 \times 10^{-13}$

9-37. $\sin\theta = 1.22\lambda/D = 1.22(600 \times 10^{-9}m)/(10 \times 10^{-2}m) = 7.32 \times 10^{-6}$

$$\approx \theta \approx 7.32 \times 10^{-6} \, radians$$

$\theta = S/R$ where $S =$ diameter of the beam on the moon and $R =$ distance to the moon.

$S = R\theta = (3.84 \times 10^8 m)(7.32 \times 10^{-6} \, radians) = 2.81 \times 10^3 \, m = 2.81 \, km$

9-41.   (a) $U_{att} = -\dfrac{ke^2}{r} = \dfrac{1.440 \, eV \cdot nm}{0.267 \, nm} = 5.39 \, eV$

(b) To form K^+ and Cl^- requires $E_{ion} = 4.34 \, eV - 3.61 \, eV = 0.73 \, eV$

$$E_d = -U_C = -\left(-\frac{ke^2}{r} + E_{ion}\right) = 5.39 \, eV - 0.73 \, eV = 4.66 \, eV$$

(c) $E_{ex} = 4.66 \, eV - 4.43 \, eV = 0.23 \, eV$ at r_0

9-45. Using the NaCl potential energy vs separation graph in Figure 9-23(b) as an example (or you can plot one using Equation 9-1):

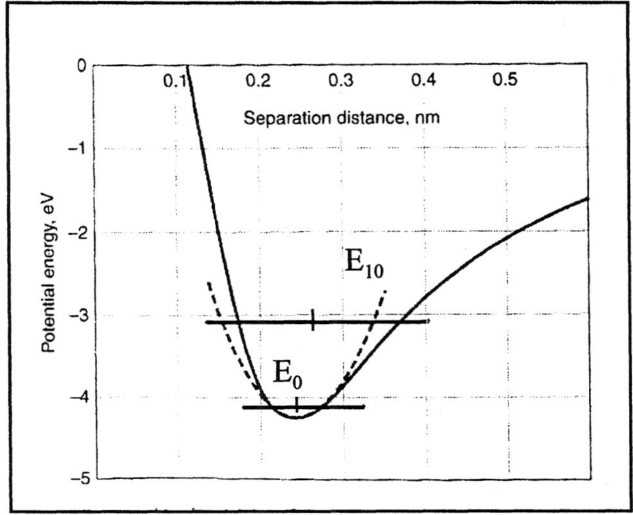

The vibrational frequency for NaCl is 1.14×10^{13} Hz (from Table 9-7) and two vibrational levels, for example $v = 0$ and $v = 10$ yield (from Equation 9-20)

$$E_0 = 1/2\, hf = 0.0236 \ eV$$

$$E_{10} = 11/2\, hf = 0.496 \ eV$$

above the bottom of the well. Clearly, the average separation for $v_{10} > v_0$.

9-49. $\quad n(E_n) = g(E_n)\, e^{-E_n/kT}$

$$\lambda_{21} = 694.3 \ nm \ \rightarrow \ E_2 - E_1 = hc/\lambda_{21} = \frac{1240 \ eV \cdot nm}{694.3 \ nm} = 1.7860 \ eV$$

$$E_2' - E_1 = 1.7860 \ eV + 0.0036 \ eV = 1.7896 \ eV$$

Where E_2 is the lower energy level of the doublet and E_2' is the upper.

Let T = 300 K, so $kT = 0.0259 \ eV$

(a) $\quad \dfrac{n(E_2')}{n(E_1)} = \dfrac{g(E_2)}{g(E_1)}\, e^{-(E_2 - E_1)/kT} = \dfrac{2}{4}\, e^{-1.7896/0.0259} = \dfrac{1}{2}\, e^{-69} = 4.91 \times 10^{-31}$

$\quad \dfrac{n(E_2)}{n(E_1)} = \dfrac{1}{2}\, e^{-1.7896/0.0259} = 5.64 \times 10^{-31}$

(Problem 9-49 continued)

(b) If only $E_2 \rightarrow E_1$ transitions produce lasing, but E_2 and E_2' are essentially equally populated, in order for population inversion between levels E_2 and E_1, at least 2/3 (rather than 1/2) of the atoms must be pumped. The required power density (see Example 9-8) is:

$$p \approx \frac{2N}{3}\left(\frac{hf}{t_s}\right) \approx \frac{2(2\times10^{19}\,cm^3)(6.63\times10^{-34}\,J\cdot s)(4.32\times10^{14}\,Hz)}{3(3\times10^{-3}s)} \approx 1273\ W/cm^3$$

9-53.   For $H^+ - H^-$ system, $U(r) = -\dfrac{ke^2}{r} + E_{ion}$

There is no E_{ex} term, the two electrons of H^- are in the $n = 1$ shell with opposite spins.

E_{ion} = ionization energy for H^+ – electron affinity for H^- = $13.6\,eV - 0.75\,eV = 12.85\,eV$.

$$U(r) = -\frac{1.440\,eV\cdot nm}{r} + 12.85\ eV \qquad \frac{dU(r)}{dr} = \frac{1.440}{r^2}$$

For $U(r)$ to have a minimum and the ionic $H^+ - H^-$ molecule to be bound, $dU/dr = 0$. As we see from the derivative, there is no non-zero or finite value of r for which this occurs.

Chapter 10 – Solid State Physics

10-1. The molar volume is $\dfrac{M}{\rho} = 2N_A r_0^3$

$$r_0 = \left[\frac{M}{2N_A\rho}\right] = \left[\frac{74.55 \; g/moles}{2(6.022\times10^{23}\,/mole)(1.984\,g/cm^3)}\right]^{1/3} = 3.15\times10^{-8} \; cm = 0.315 \; nm$$

10-5. Cohesive energy (LiBr) $= 788\times10^3\,J/mol\left(\dfrac{1\,mol}{6.02\times10^{23}\;ion\;pairs}\right)\left(\dfrac{1\,eV}{1.60\times10^{-19}}\right)$

$$= 8.182 \; eV/ion\;pair = 4.09 \; eV/atom$$

This is about 32% larger than the value in Table 10-1.

10-9. (a) $\quad \rho = \dfrac{m_e\langle v\rangle}{ne^2\lambda} \qquad$ (Equation 10-17)

$$= \frac{(9.11\times10^{-31}\,kg)(1.17\times10^5\,m/s)}{(8.47\times10^{28}\,electrons/m^3)(1.60\times10^{-19}\,C)^2(0.4\times10^{-9}\,m)}$$

$$= 1.23\times10^{-7} \; \Omega\cdot m$$

(b) $\langle v\rangle \propto \left(kT/m_e\right)^{1/2} \quad$ (from Equation 10-9)

$$\langle v\rangle_{100} = \left(\frac{100\,K}{300\,K}\right)^{1/2} = \frac{1}{\sqrt{3}}$$

$$\rho_{100} = \rho_{300}/\sqrt{3} = 7.00\times10^{-8} \; \Omega\cdot m$$

10-13. (a) $n = 2\rho N_A / M$ for two free electrons/atom

$$n = \frac{2\left(1.74\,g/cm^3\right)\left(6.022 \times 10^{23}\,/mole\right)}{24.31\,g/mole} = 8.62 \times 10^{22}\,/cm^3 = 8.62 \times 10^{28}\,/m^3$$

(b) $n = \dfrac{2\left(7.1\,g/cm^3\right)\left(6.022 \times 10^{23}\,/mole\right)}{65.37\,g/mole} = 13.1 \times 10^{22}\,/cm^3 = 13.1 \times 10^{28}\,/m^3$

Both are in good agreement with the values in Table 10-3, $8.61 \times 10^{28}\,/m^3$ for Mg and $13.2 \times 10^{28}\,/m^3$ for Zn.

10-17. (a) For Ag: $E_F = \dfrac{h^2}{2m}\left(\dfrac{3N}{8\pi V}\right)^{2/3} = \dfrac{\left(1240\,eV \cdot nm \times 10^{-9}\,m/nm\right)^2}{2\left(0.511 \times 10^6\,eV\right)}\left(\dfrac{3 \times 5.86 \times 10^{28}\,m^{-3}}{8\pi}\right)^{2/3} = 5.50\,eV$

For Fe: Similarly, $E_F = 11.2$ eV

(b) For Ag: $E_F = kT_F$ (Equation 10-38)

$$T_F = \frac{E_F}{k} = \frac{5.50\,eV}{8.617 \times 10^{-5}\,eV/K} = 6.38 \times 10^4\,K$$

For Fe: Similarly, $T_F = 13.0 \times 10^4$ K

Both results are in close agreement with the values given in Table 10-3.

10-21. $C_v(electrons) = \dfrac{\pi^2}{2}R\dfrac{T}{T_F}$ (Equation 10-45)

$C_v(electrons) = \dfrac{\pi^2}{2}R\dfrac{kT}{E_F}$ because $E_F = kT_F$. C_v due to the lattice vibrations is 3R, assuming

$T \gg T_E$ (rule of Dulong and Petit): $\dfrac{\pi^2}{2}R\dfrac{kT}{E_F} = 0.10(3R)$

$$T = \frac{0.10\,(3)(2)E_F}{\pi^2 k} = \frac{(0.60)\,(7.06\,eV)}{\pi^2\left(8.617 \times 10^{-5}\,eV/K\right)} = 4.98 \times 10^3\,K$$

This temperature is much higher than the Einstein temperature for a metal such as copper.

10-25.

$$\chi = \frac{\mu_0 M}{B} = \frac{\mu_0 \rho \mu^2}{kT}$$

$$\chi \text{ units} = \left(\frac{N}{A^2}\right)\left(\frac{1}{m^3}\right)\left(\frac{J}{T}\right)^2\left(\frac{1}{J}\right)$$

$$= \frac{NJ^2}{A^2 m^3 T^2 J}$$

$$= \frac{NJ}{A^2 m^3 (Wb/m)^2}$$

$$= \frac{NJ}{A^2 m^3 (N/Am)^2} = \frac{NJA^2 m^2}{A^2 m^3 N^2}$$

$$= \frac{J}{Nm} = \frac{Nm}{Nm} = 1 \quad \text{dimensionless}$$

10-29. (a) $N = \dfrac{mN_A}{M} = \dfrac{\rho V N_A}{M} = \dfrac{(2.33\,g/cm^3)(100\,nm \times 10^{-7}\,cm/nm)^3 (6.02 \times 10^{23}/mol)}{28\,g/mol}$

$$= 5.01 \times 10^7 \; Si \; atoms$$

(b) $\quad \Delta E \approx 13\,eV/(4 \times 5.01 \times 10^7) = 6.5 \times 10^{-8}\,eV$

10-33. $\quad E = kT = 0.01\,eV \quad \therefore \; T = 0.01\,eV/8.617 \times 10^{-5}\,eV/K = 116K$

10-37. $\quad I_{net} = I_0\left(e^{eV_b/kT} - 1\right) \qquad$ (Equation 10-64)

Assuming T = 300K,

$$\frac{I(0.2\,V) - V(0.1\,V)}{I(0.1\,V)} = \frac{I_0\left(e^{e(0.2\,V)/kT} - 1\right) - I_0\left(e^{e(0.1\,V)/kT} - 1\right)}{I_0\left(e^{e(0.1\,V)/kT} - 1\right)}$$

$$= \frac{e^{e(0.2\,V)/kT} - e^{e(0.1\,V)/kT}}{e^{e(0.1\,V)/kT} - 1}$$

$$= 47.6/1$$

10-41.   (a) $E_g = 3.5\,kT_c$ For *Sn*: $T_c = 3.722$ K

$$E_g = 3.5\left(8.617 \times 10^{-5}\,eV/K\right)\left(3.722\,K\right) = 0.0011\ eV$$

This is about twice the measured value.

(b) $E_g = hc/\lambda$

$$\lambda = hc/E = 1240\,eV \cdot nm / 6 \times 10^{-4}\,eV$$

$$= 2.07 \times 10^6\,nm = 2.07 \times 10^{-3}\ m$$

10-45.

$-e$	$+e$	$-e$	$+e$	$-e$	$+e$	$-e$	$+e$	$-e$	$+e$	$-e$	$+e$	$-e$
•	•	•	•	•	•	•	•	•	•	•	•	•
-6	-5	-4	-3	-2	-1	0	1	2	3	4	5	6

$$\rightarrow\!|\ r_0\ |\!\leftarrow$$

(a) For the negative ion at the origin (position 0) the attractive potential energy is:

$$V = -\frac{2ke^2}{r_0}\left(1 - \frac{1}{2} + \frac{1}{3} - \frac{1}{4} + \frac{1}{5} - \frac{1}{6} + \cdots\right)$$

(b) $V = -\alpha\dfrac{ke^2}{r_0}$, so the Madelung constant is

$$\alpha = 2\left(1 - \frac{1}{2} + \frac{1}{3} - \frac{1}{4} + \frac{1}{5} - \frac{1}{6} + \cdots\right)$$

Noting that $\ln(1+x) = x - \dfrac{x^2}{2} + \dfrac{x^3}{3} - \dfrac{x^4}{4} + \cdots$, $\quad \ln(2) = 1 - \dfrac{1}{2} + \dfrac{1}{3} - \dfrac{1}{4} + \cdots$

and $\alpha = 2\,\ln 2 = 1.386$

10-49.

$$\lambda = \frac{m_{\text{eff}} \langle v \rangle}{n \rho e^2} \quad \text{(Equation 10-17)}$$

$$\langle v \rangle = (3kT/m_{\text{eff}})^{1/2} = \left[\frac{3(1.38 \times 10^{-23}\,J/K)(300k)}{0.2\,(9.11 \times 10^{-31}\,kg)} \right]^{1/2}$$

$$\langle v \rangle = 2.61 \times 10^5\,m/s$$

Substituting into λ:

$$\lambda = \frac{0.2\,(9.11 \times 10^{-31}\,kg)(2.61 \times 10^5\,m/s)}{(10^{-22}\,m^{-3})(5 \times 10^{-3}\,\Omega m)(1.60 \times 10^{-19}\,C)^2}$$

$$\lambda = 3.7 \times 10^{-8}\,m = 37\,nm$$

For Cu: $u_F = (2E_F/m_e)^{1/2} = \left[\dfrac{2(7.06\,eV)}{9.11 \times 10^{-31}\,kg} \right]^{1/2}$

$$u_F = 1.57 \times 10^6\,m/s$$

$$n = 8.47 \times 10^{28}\,m^{-3} \quad \text{and} \quad \rho = 1.7 \times 10^{-8}\,\Omega \cdot m \qquad \text{(Example 10-6)}$$

Substituting as above, $\lambda = 3.9 \times 10^{-8}$ m $= 39$ nm

The mean free paths are approximately equal.

10-53. (a) The modified Schrödinger equation is:

$$-\frac{\hbar^2}{2m^* r^2}\frac{d}{dr}\left(r^2 \frac{dR(r)}{dr} \right) + \left[-\frac{ke^2}{r\kappa} + \frac{\hbar^2 \ell(\ell+1)}{2m^* r^2} \right] R(r) = ER(r)$$

The solution of this equation, as indicated following Equation 7-24, leads to solutions

of the form: $R_{n\ell}(r) = a_o' e^{-r/a_0' n} r^{-1} \mathcal{L}_{n\ell}(r/a_0')$, where $a_0' = (\hbar^2 \kappa^2/ke^2 m^*)$.

(b) By substitution into Equation 7-25, the allowed energies are:

$$E_n = -\frac{1}{2}\left(\frac{ke^2}{\hbar\kappa} \right)^2 \frac{m^*}{n^2} = -\frac{E_1}{n^2} \quad \text{where} \quad E_1 = \frac{1}{2}\left(\frac{ke^2}{l\kappa} \right)^2 m^*$$

(Problem 10-53 continued)

(c) For *As* electrons in *Si* $m^* = 0.2m_c$ (see Problem 10-30) and $\kappa(Si) = 11.8$,

$$E_1 = -\frac{1}{2}\frac{\left[\left(9\times10^9\,N\cdot m^2/C^2\right)\left(1.60\times10^{-19}\,C\right)^2\right]^2}{\left(1.055\times10^{-34}\,J\cdot s\right)^2}\cdot\frac{(0.2)\left(9.11\times10^{-31}\,kg\right)}{(11.8)^2}$$

$$= -3.12\times10^{-21}\,J = -0.0195\ eV$$

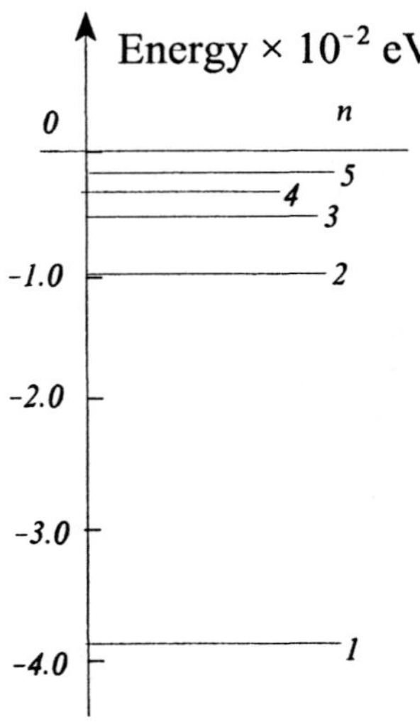

10-57. (a) $\quad M = \mu\left(\rho_+ - \rho_-\right) \quad\Rightarrow\quad \dfrac{M}{\rho} = \mu\dfrac{\left(\rho_+ - \rho_-\right)}{\rho}$

$$\frac{M}{\rho} = \mu\frac{e^{\mu B/kT} - e^{-\mu B/kT}}{e^{\mu B/kT} + e^{-\mu B/kT}} = \mu\tanh\left(\mu B/kT\right)$$

and

$$M = \mu\rho\tanh\left(\mu B/kT\right)$$

(b) For $\mu B \ll kT$, $T\gg0$ and $\tanh(\mu B/kT) \approx \mu B/kT$

$$\chi = \frac{\mu_0 M}{B} = \frac{\mu_0\mu\rho\mu B}{BkT} = \frac{\mu_0\rho\mu^2}{kT}$$

Chapter 11 – Nuclear Physics

11-1.

Isotope	Protons	Neutrons
^{18}F	9	9
^{25}Na	11	14
^{51}V	23	28
^{84}Kr	36	48
^{120}Te	52	68
^{148}Dy	66	82
^{175}W	74	101
^{222}Rn	86	136

11-5. The two proton spins would be antiparallel in the ground state with $S = 1/2 - 1/2 = 0$. So the deuteron spin would be due to the electron and equal to $1/2\hbar$. Similarly, the proton magnetic moments would add to zero and the deuteron's magnetic moment would be 1 μ_B. From Table 11-1, the observed deuteron spin is $1\hbar$ (rather than $1/2\hbar$ found above) and the magnetic moment is 0.857 μ_N, about 2000 times smaller than the value predicted by the proton-electron model.

11-9. $B = ZM_Hc^2 + Nm_Nc^2 - M_Ac^2$ (Equation 11-11)

(a) $^{9}_{4}Be_5$

$$B = 4(1.007825\,uc^2) + 5(1.008665\,uc^2) - 9.012182\,uc^2$$

$$= 0.062443\,uc^2 = (0.062443\,uc^2)(931.5\,MeV/uc^2)$$

$$= 58.2\,MeV$$

$$B/A = 58.2\,MeV/9\,nucleons = 6.46\,MeV/nucleon$$

(b) $^{13}_{6}C_7$

$$B = 6(1.007825\,uc^2) + 7(1.008665\,uc^2) - 13.003355\,uc^2$$

$$= 0.104250\,uc^2 = (0.104250\,uc^2)(931.5\,MeV/uc^2)$$

$$= 91.1\,MeV$$

$$B/A = 91.1\,MeV/13\,nucleons = 7.47\,MeV/nucleon$$

(Problem 11-9 continued)

(c) $^{57}_{26}Fe_{31}$ $\quad B = 26(1.007825\,uc^2) + 31(1.008665\,uc^2) - 56.935396\,uc^2$

$$= 0.536669\,uc^2 = (0.536669\,uc^2)(931.5\,MeV/uc^2)$$

$$= 499.9\,MeV$$

$$B/A = 499.9\,MeV/57\,nucleons = 8.77\,MeV/nucleon$$

11-13. $R = (1.07 \pm 0.02)A^{1/3}\,fm$ (Equation 11-5) $\quad R = 1.4A^{1/3}\,fm$ (Equation 11-6)

(a) ^{16}O: $R = 1.07A^{1/3} = 2.70\,fm$ $\quad and \quad R = 1.4A^{1/3} = 3.53\,fm$

(b) ^{63}Cu: $R = 1.07A^{1/3} = 4.26\,fm$ $\quad and \quad R = 1.4A^{1/3} = 5.57\,fm$

(c) ^{208}Pb: $R = 1.07A^{1/3} = 6.34\,fm$ $\quad and \quad R = 1.4A^{1/3} = 8.30\,fm$

$$\Delta U = \frac{3}{5}\frac{1}{4\pi\epsilon_0}\frac{e^2}{R}\left(Z^2 - (Z-1)^2\right)$$

11-17. $R = R_0 e^{-\lambda t} = R_0 e^{-(\ln 2)t/t_{1/2}}$ (Equation 11-19)

(a) At $t = 0$: $R = R_0 = 115.0\,decays/s$

At $t = 2.25\,h$: $R = 85.2\,decays/s$

$$85.2\,decays/s = (115.0\,decays/s)\,e^{-\lambda(2.25h)}$$

$$(85.2/115.0) = e^{-\lambda(2.25)}$$

$$\ln(85.2/115.0) = -\lambda(2.25h)$$

$$\lambda = -\ln(85.2/115.0)/2.25\,h = 0.133\,h^{-1}$$

$$t_{1/2} = \ln 2/\lambda = \ln 2/0.133\,h^{-1} = 5.21\,h$$

(b) $\quad \left|\dfrac{dN}{dt}\right| = \lambda N \quad \rightarrow \quad \left|\dfrac{dN_0}{dt}\right| = R_0 = \lambda N_0$ (from Equation 11-17)

$$N_0 = R_0/\lambda = (115.0\,atoms/s)/(0.133\,h^{-1})(1\,h/3600\,s)$$

$$= 3.11 \times 10^6\,atoms$$

11-21. ^{62}Cu is produced at a constant rate R_0, so the number of ^{62}Cu atoms present is:

$N = R_0/\lambda\left(1 - e^{-\lambda t}\right)$ (from Equation 11-26). Assuming there were no ^{62}Cu atoms initially

present. The maximum value N can have is $R_0/\lambda = N_0$,

$$N = N_0\left(1 - e^{-\lambda t}\right)$$

$$0.90\,N_0 = N_0\left(1 - e^{-t\ln2/t_{1/2}}\right)$$

$$e^{-t\ln2/t_{1/2}} = 1 - 0.90 = 0.10$$

$$-t\ln2/t_{1/2} = \ln(0.10)$$

$$t = -10\ln(0.10)/\ln2 = 33.2 \text{ min}$$

11-25. $\log t_{1/2} = A E_\alpha^{-1/2} + B$ (Equation 11-18)

$\left.\begin{array}{l} \text{for } t_{1/2} = 10^{10}\,s,\, E_\alpha = 5.4\,MeV \\[2mm] \text{for } t_{1/2} = 1\,s,\, E_\alpha = 7.0\,MeV \end{array}\right\}$ ← from Figure 11-16

$\log 10^{10} = A(5.4)^{-1/2} + B$
  (i) $10 = 0.4303\,A + B$
      $\log 1 = A(7.0)^{-1/2} + B$
  (ii) $0 = 0.3780\,A + B$ → $B = -0.3780\,A$
Substituting (ii) into (i),
    $10 = 0.4303\,A - 0.3780\,A = 0.0523\,A$, $A = 191$, $B = -0.3780\,A = -72.2$

11-29. $^{67}Ga \xrightarrow{E.C.} {}^{67}Zn + \nu_e$

$$Q = M\left(^{67}Ga\right)c^2 - M\left(^{67}Zn\right)c^2$$

$$= 66.9282\,uc^2 - 66.927129\,uc^2$$

$$= 0.001075\,uc^2\left(931.50\,MeV/uc^2\right) = 1.00\,MeV$$

11-33. $^8Be \rightarrow 2\alpha$

$$Q = M(^8Be)c^2 - M(^4He)c^2$$

$$= 8.005304\,uc^2 - 2(4.002602)uc^2$$

$$= 0.000100\,uc^2\,(931.50\,MeV/uc^2) = 0.093\,MeV = 93\,keV$$

Thus, the lower energy configuration for 4 protons and 4 neutrons is two α particles rather than one ^{8}Be.

11-37. The range R of a force mediated by an exchange particle of mass m is:

$$R = \hbar/mc \quad \text{(Equation 11-50)}$$

$$mc^2 = \hbar c/R = 197.3\,MeV{\cdot}fm\,/\,0.25\,fm = 789\,MeV$$

$$m = 789\,MeV/c^2$$

11-41. ^{36}S, ^{53}Mn, ^{82}Ge, ^{88}Sr, ^{94}Ru, ^{131}In, ^{145}Eu

11-45.

$$^{30}_{14}Si \qquad j = 0$$

$$^{37}_{17}Cl \qquad j = 3/2$$

$$^{55}_{27}Co \qquad j = 7/2$$

$$^{90}_{40}Zr \qquad j = 0$$

$$^{107}_{49}In \qquad j = 9/2$$

11-49. The number N of ^{144}Nd atoms is:

$$N = \frac{53.94\,g\left(6.02\times10^{23}\,atoms/mol\right)}{144\,g/mol} = 2.25\times10^{23}\,atoms$$

$$-\frac{dN}{dt} = \lambda N \quad\rightarrow\quad \lambda = (-dN/dt)/N$$

$$= \left(2.36\,s^{-1}\right)/\left(2.25\times10^{23}\right)$$

$$= 1.05\times10^{-23}\,s^{-1}$$

$$t_{1/2} = \ln 2 / \lambda = \ln 2 / 1.05\times10^{-23}\,s^{-1} = 6.61\times10^{22}\,s = 2.09\times10^{15}\,y$$

11-53. $B = ZM\left(^1H\right)c^2 + Nm_n c^2 - M_A c^2$

For ^{3}He: $B = 2\left(1.007825\,uc^2\right) + 1.008665\,uc^2 - 3.016029\,uc^2$

$$= 0.008826\,uc^2\left(931.50\,MeV/uc^2\right) = 7.72\,MeV$$

For ^{3}H: $B = 1.007825\,uc^2 + 2\left(1.008665\,uc^2\right) - 3.016049\,uc^2$

$$= 0.009106\,uc^2\left(931.50\,MeV/uc^2\right) = 8.48\,MeV$$

$R = R_0 A^{1/3} = 1.2\,fm\,3^{1/3} = 1.730\,fm = 1.730\times10^{-15}\,m$

$u_c = ke^2/R = ke/R(eV) = 8.32\times10^5\,eV = 0.832\,MeV$ or about 1/10 of the binding energy.

11-57. (a) Using $\partial M/\partial Z = 0$ from Problem 11-44,

$$Z = \frac{\left(m_n - m_p\right) + 4a_4}{2a_3 A^{-1/3} + 8a_4 A^{-1}} \quad \text{where } a_3 = 0.75\,MeV/c^2 \,, a_4 = 93.2\,MeV/c^2$$

$$= \frac{1 + \left(m_n - m_p\right)/4a_4}{2A^{-1} + \left(a_3 A^{-1/3}/2a_4\right)} = \frac{A}{2}\frac{\left[1 + \left(m_n - m_p\right)/4a_4\right]}{\left[1 + a_3 A^{2/3}/4a_4\right]}$$

(b) & (c) For A = 29: $Z = \dfrac{29}{2}\dfrac{\left[1 + (1.008665 - 1.007276)(931.5)/(4)(93.2)\right]}{\left[1 + 0.75(29)^{2/3}/(4)(93.2)\right]} = 14$

The only stable isotope with A = 29 is $^{29}_{14}Si$

For A = 59: Computing as above with A = 59 yields Z = 29. The only stable isotope with A = 59 is $^{59}_{27}Co$

For A = 78: Computing as above with A = 78 yields Z = 38. $^{78}_{38}Sr$ is not stable. Stable isotopes with A = 78 are $^{78}_{34}Se$ and $^{78}_{36}Kr$.

For A = 119: Computing as above with A = 119 yields Z = 59. $^{119}_{59}Pr$ is not stable.

The only stable isotope with A = 119 is $^{119}_{50}Sn$

For A = 140: Computing as above with A = 140 yields Z = 69. $^{140}_{69}Tm$ is not stable. The

only stable isotope with A = 140 is $^{140}_{58}Ce$

The method of finding the minimum Z for each A works well for $A \le 60$, but deviates increasingly at higher A values.

11-61. (a) If the electron's kinetic energy is 0.782 MeV, then its total energy is:

$$E = 0.782\,MeV + m_e c^2 = 0.782\,MeV + 0.511\,MeV = 1.293\,MeV$$

$$E^2 = (pc)^2 + \left(m_e c^2\right)^2 \quad \text{(Equation 2-32)}$$

$$p = \left(E^2 - \left(m_e c^2\right)^2\right)^{1/2}/c$$

$$= \left[(1.293\,MeV)^2 - (0.511\,MeV)^2\right]^{1/2}/c$$

$$= 1.189\,MeV/c$$

(b) For the proton $p = 1.189\,MeV/c$ also, so

(Problem 11-61 continued)

$$E_{kin} = p^2/2m = (pc)^2/2mc^2$$

$$= (1.189\,MeV)^2/(2)(938.28\,MeV) = 7.53 \times 10^{-4}\,MeV = 0.753\,keV$$

(c) $\dfrac{7.53 \times 10^{-4}\,MeV}{0.782\,MeV} \times 100 = 0.0963\,\%$

Chapter 12 – Nuclear Reactions and Applications

12-1. (a) $Q = M(^2H)c^2 + M(^2H)c^2 - M(^3H)c^2 - M(^1H)c^2$

$= 2(2.014102\,uc^2) - 3.016049\,uc^2 - 1.007825\,uc^2$

$= 0.004330\,uc^2\,(931.5\,MeV/uc^2) = 4.03\,MeV$

(b) $Q = M(^3He)c^2 + M(^2H)c^2 - M(^4He)c^2 - M(^1H)c^2$

$= 3.016029\,uc^2 + 2.014102\,uc^2 - 4.002602\,uc^2 - 1.007825\,uc^2$

$= 0.019704\,uc^2\,(931.5\,MeV/uc^2) = 18.35\,MeV$

(c) $Q = M(^6Li)c^2 + m_n c^2 - M(^3H)c^2 - M(^4He)c^2$

$= 6.01512\,uc^2 + 1.008665\,uc^2 - 3.016049\,uc^2 - 4.002602\,uc^2$

$= 0.005135\,uc^2\,(931.5\,MeV/uc^2) = 4.78\,MeV$

12-5. The number of ^{75}As atoms in sample N is:

$$N = \frac{V\rho N_A}{M} = \frac{(1\,cm \times 2\,cm \times 30\,\mu m \times 10^{-4}\,cm/\mu m)(5.73\,g/cm^3)(6.02 \times 10^{23}\,atoms/mol)}{74.9216\,g/mol}$$

$= 2.76 \times 10^{20}\quad {}^{75}As\ atoms$

The reaction rate R per second per ^{75}As atom is:

$R = \sigma I$ (Equation 12-5)

$= (4.5 \times 10^{-24}\,cm^2/{}^{75}As)(0.95 \times 10^{13}\,neutrons/cm^2 \cdot s)$

$= 4.28 \times 10^{-11}\,s^{-1}$

Reaction rate $= NR$

$= (2.76 \times 10^{20}\,atoms)(4.28 \times 10^{-11}/s \cdot atom)$

$= 1.18 \times 10^{10}/s$

12-9.
$$P = \frac{dW}{dt} = E\frac{dN}{dt}$$

$$\frac{dN}{dt} = \frac{P}{E} = \frac{500 \times 10^6 \, J/s}{200 \times 10^6 \, eV/fission}\left(\frac{1 \, eV}{1.60 \times 10^{-19} \, J}\right)$$

$$= 1.56 \times 10^{19} \, fissions/s$$

12-13. The reactions per ^{238}U atom R is:

$$R = \sigma I \qquad (\text{Equation } 12\text{-}5)$$

$$= \left(0.02 \times 10^{-24} \, cm^2/atom\right)\left(5.0 \times 10^{11} \, n/m^2\right)\left(\frac{1 \, m^2}{10^4 \, cm^2}\right) = 1.00 \times 10^{-18}/atom$$

The number N of ^{238}U atoms is:

$$N = \frac{(5.0 \, g)\left(6.02 \times 10^{23} \, atoms/mol\right)}{238.051 \, g/mol} = 1.26 \times 10^{22} \quad ^{238}U \text{ atoms}$$

Total ^{239}U atoms produced $= R \, N$

$$= \left(1.00 \times 10^{-18}/atom\right)\left(1.26 \times 10^{22} \, atoms\right) = 1.26 \times 10^4 \quad ^{239}U \text{ atoms}$$

12-17 (a) $Q = M(^{235}U)c^2 + m_n c^2 - M(^{120}Cd)c^2 - M(^{110}Ru)c^2 + 5 m_n c^2$

$$= 235.043924 uc^2 + 1.008665 uc^2 - 119.909851 uc^2 - 109.913859 uc^2$$

$$- 5\left(1.008665 \, uc^2\right)$$

$$= 1.186 \, uc^2\left(931.50 \, MeV/uc^2\right) = 1.10 \times 10^3 \, MeV$$

(b) This reaction is not likely to occur. Both product nuclei are neutron-rich and highly unstable.

12-21. The number of X rays counted during the experiment equals the number of atoms of interest in the same times the cross section for activation σ_x times the particle beam intensity, where

I = proton intensity = 650 nA $\times$ (e C/proton)$^{-1}$; $\sigma_x = 650$ b $= 650 \times 10^{-24}$ cm^2

m = mass = 0.35 mg/cm$^2 \times 0.00001$; n = number of atoms of interest = mN_A/A

t = exposure time; detector efficiency = 0.0035

ϵ = overall efficiency = 0.60 $\times$ detector efficiency

$$N = I\sigma_x \frac{mN_A}{A} t\epsilon$$

$$N = \left(\frac{650 \times 10^{-9}\, C/s}{1.60 \times 10^{-19}\, C/proton} \right) \left(650 \times 10^{-24}\, cm^2 \right) \left(\frac{(0.35 \times 10^{-3}\, g/cm^2)(0.00001)(6.02 \times 10^{23}\, mol^{-1})}{80\, g/mol} \right)$$

$$\times\ (15\,min \times 60\,s/min)(0.60 \times 0.0035)$$

$$N = 1.31 \times 10^5 \text{ counts in 15 minutes}$$

12-25. If from live wood, the decay rate would be 15.6 disintegrations/g·min. The actual rate is 2.05 disintegrations/g·min.

$$\left(\frac{1}{2} \right)^n = \frac{2.05 \text{ decays/g} \cdot \text{min}}{15.6 \text{ decays/g} \cdot \text{min}} \qquad (\text{from Example 12-13})$$

$$2^n = 15.6 / 2.05$$

$$n\ln 2 = \ln(15.6 / 2.05)$$

$$n = \ln(15.6/2.05)/\ln 2 = 2.928 \text{ half lives of } {}^{14}C$$

$$\text{Age of spear thrower} = nt_{1/2} = (2.928)(5730 y) = 16,800 y$$

12-29. $Q = 200$ MeV/fission. $E = NQ = 7.0 \times 10^{19}\, J = N(200\, MeV/fission)(1.60 \times 10^{-13}\, J/MeV)$

$$N = \frac{7.0 \times 10^{19}\, J}{(200\, MeV/fission)(1.60 \times 10^{-13}\, J/MeV)} = 2.19 \times 10^{30}\, fissions$$

Number of moles of ^{235}U needed $= N/N_A = 2.19 \times 10^{30}/6.02 \times 10^{23} = 3.63 \times 10^6\, moles$

Fissioned mass/y $= (3.63 \times 10^6\, moles)(235\, g/mole) = 8.54 \times 10^8\, g = 8.54 \times 10^5\, kg$

(Problem 12-29 continued)

This is 3% of the mass of ^{235}U atoms needed to produce the energy consumed.

Mass needed to produce $7.0 \times 10^{19} J = 8.54 \times 10^5 \, kg/0.03 = 2.85 \times 10^7 \, kg$.

Since the energy conversion system is 25% efficient:

Total mass of ^{235}U needed $= 1.14 \times 10^8 \, kg$.

12-33. The net reaction is: $5 \, ^2H \rightarrow \, ^3He + \, ^4He + \, ^1H + n + 25 \, MeV$

Energy release $/ \, ^2H = 5 \, MeV$ (assumes equal probabilities)

4ℓ water $\rightarrow 4000 \, g / [2(1.007825) + 15.994915] \, g/mol = 222.1 \, moles$

4ℓ water thus contains $2(222.1)$ moles of hydrogen, of which 1.5×10^{-4} is ^2H, or

Number of ^2H atoms $= [2(222.1) \, moles](6.02 \times 10^{23} \, atoms/mole)(1.5 \times 10^{-4}) = 4.01 \times 10^{22}$

Total energy release $= (4.01 \times 10^{22}) 5 \, MeV = 2.01 \times 10^{23} \, MeV = 3.22 \times 10^{10} \, J$

Because the U.S. consumes about $7.0 \times 10^{19} \, J/y$, the complete fusion of the ^2H in 4ℓ of water

would supply the nation for about $1.45 \times 10^{-2} s = 14.5 \, ms$

12-37. At this energy, neither particle is relativistic, so

$$E_{He} = \frac{p_{He}^2}{2 m_{He}} \qquad E_n = \frac{p_n^2}{2 m_n} \qquad p_{He} = p_n \qquad E_{He} + E_n = 17.7 \, MeV$$

$$2 m_{He} E_{He} = 2 m_n E_n = 2 m_n (17.7 \, MeV - E_{He})$$

$$(m_{He} + m_n) E_{He} = 17.7 \, MeV \, m_n \qquad \text{Therefore, } E_{He} = \frac{m_n}{m_{He} + m_n} 17.7 \, MeV$$

$$E_{He} = \frac{1.008665 \, u (17.7 \, MeV)}{4.002602 \, u + 1.008665 \, u} = 3.56 \, MeV$$

$$E_n = 17.7 \, MeV - E_{He} = (17.7 - 3.56) MeV = 14.1 \, MeV$$

Chapter 13 – Particle Physics

13-1. (a) Because the two pions are initially at rest, the net momentum of the system is zero, both before and after annihilation. For the momentum of the system to be zero after the interaction, the momenta of the two photons must be equal in magnitude and opposite in direction, i.e., their momentum vectors must add to zero. Because the photon energy is $E = pc$, their energies are also equal.

   (b) The energy of each photon equals the rest energy of a π^+ or a π^-.

   $$E = m_\pi c^2 = 139.6\ MeV \quad \text{(from Table 13-1)}$$

   (c) $E = hf = hc/\lambda$ Thus, $\lambda = \dfrac{hc}{E} = \dfrac{1240\ MeV \cdot fm}{139.6\ MeV} = 8.88\ fm$

13-5. (a) A single photon cannot conserve both energy and momentum.

   (b) To conserve momentum each photon must have equal and opposite momenta so that the total momentum is zero. Thus, they have equal energies, each equal to the rest energy of a proton: $E_\gamma = m_p c^2 = 938.28\ MeV$

   (c) $E_\gamma = h\nu = hc/\lambda$ $\therefore$ $\lambda = \dfrac{hc}{E_\gamma} = \dfrac{1240\ MeV \cdot fm}{938.28\ MeV} = 1.32\ fm$

   (d) $\nu = \dfrac{c}{\lambda} = \dfrac{3.00 \times 10^8\ m/s}{1.32 \times 10^{-15}\ m} = 2.27 \times 10^{23}\ Hz$

13-9. $\pi^0 \rightarrow \gamma + \gamma$ is caused by the electromagnetic interaction; $\pi^- \rightarrow \mu^- + \bar{\nu}_\mu$ is caused by the weak interaction. The electromagnetic interaction is the faster and stronger, so the π^0 will decay more quickly; the π^- will live longer.

13-13. (a)

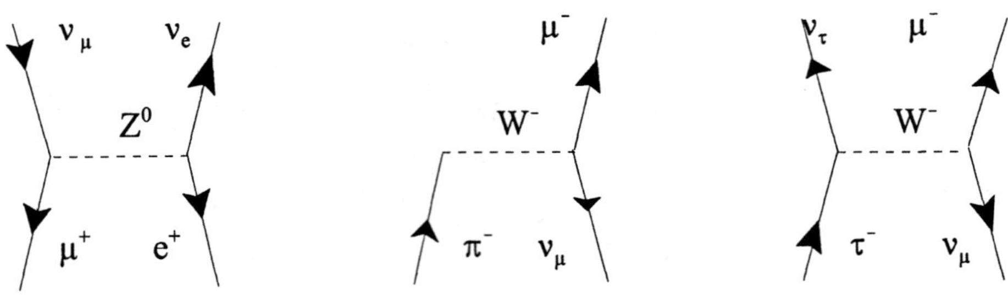

13-17. (a) $n + n$ $\qquad T_3 = -\dfrac{1}{2} - \dfrac{1}{2} = -1$ $\qquad T = 1$

(b) $n + p$ $\qquad T_3 = -\dfrac{1}{2} + \dfrac{1}{2} = 0$ $\qquad T = 1 \; or \; 0$

(c) $\pi^+ + p$ $\qquad T_3 = 1 + \dfrac{1}{2} = \dfrac{3}{2}$ $\qquad T = \dfrac{3}{2}$

(d) $\pi^- + n$ $\qquad T_3 = -1 - \dfrac{1}{2} = -\dfrac{3}{2}$ $\qquad T = \dfrac{3}{2}$

(e) $\pi^+ + n$ $\qquad T_3 = 1 - \dfrac{1}{2} = \dfrac{1}{2}$ $\qquad T = \dfrac{1}{2} \; or \; \dfrac{3}{2}$

13-21. (a) $0 + 0 \rightarrow 0 + 0$ $\qquad$ S is conserved.

(b) $-2 \rightarrow 0 - 1$ $\qquad$ S is not conserved.

(c) $-1 \rightarrow -1 + 0$ $\qquad$ S is conserved.

(d) $0 + 0 \rightarrow 0 - 1$ $\qquad$ S is not conserved.

(e) $-3 \rightarrow -2 + 0$ $\qquad$ S is not conserved.

13-25. (a) $T_3 = 0$ (from Figure 13-22a)

(b) $T = 1$ or 0 just as for ordinary spin.

(c) uds $B = 1/3 + 1/3 + 1/3 = 1$ $C = 2/3 - 1/3 - 1/3 = 0$

$S = 0 + 0 + -1 = -1$ The $T = 1$ state is the Σ^0.   The $T = 0$ state is the Λ^0.

13-29.

$\Lambda^0 \rightarrow \quad p \quad + \quad \pi^-$
$(uds) \quad (uud) \quad (\bar{u}d)$

Weak decay

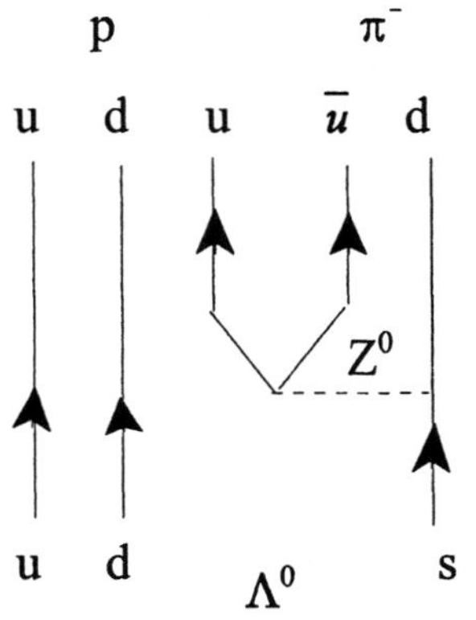

13-33. (a) Being a meson, the D^+ is constructed of a quark-antiquark pair. The only combination with *charge* $= +1$, *charm* $= +1$, and *strangeness* $= 0$ is the $c\bar{d}$. (See Table 13-7.)

(b) The D^-, antiparticle of the D^+, has the quark structure $\bar{c}d$.

13-37. (a) $p \rightarrow e^+ + \Lambda^0 + \nu_e$

$$Q = \left(m_p c^2 - M(\Lambda^0)c^2 - m_e c^2\right) MeV$$

$$= (938.3 - 1116 - 0.511) MeV = -178\, MeV$$

Energy is not conserved.

(b) $p \rightarrow \pi^+ + \gamma$

Spin (angular momentum) $\dfrac{1}{2} \rightarrow 0 + 1 = 1$. Angular momentum is not conserved.

(Problem 13-37 continued)

  (c) $p \to \pi^+ + K^0$

  Spin (angular momentum) $\frac{1}{2} \to 0 + 0 = 0$. Angular momentum is not conserved.

13-41. (a) The final products (p, γ, e$^-$, neutrinos) are all stable.

  (b) $\Xi^0 \to p + e^- + \bar{\nu}_e + \bar{\nu}_\mu + \nu_\mu$

  (c) Conservation of charge: $0 \to +1 - 1 + 0 + 0 + 0 = 0$

  Conservation of baryon number: $1 \to 1 + 0 + 0 + 0 + 0 = 1$

  Conservation of lepton number:

        (i) for electrons: $0 \to 0 + 1 - 1 + 0 + 0 = 0$

        (ii) for muons: $0 \to 0 + 0 + 0 - 1 + 1 = 0$

  Conservation of strangeness: $-2 \to 0 + 0 + 0 + 0 + 0 = 0$

        Even though the chain has $\Delta S = +2$, no individual reaction in the chain exceeds $\Delta S = +1$, so they can proceed via the weak interaction.

  (d) No, because energy is not conserved.

13-45. (a) The decay products in the chain are not all stable. In particular, the neutron decays via

  $n \to p + e^- + \bar{\nu}_e$

  (b) The net effect of the chain reaction is:

  $\Omega^- \to p + 3e^- + e^+ + 3\bar{\nu}_e + 2\bar{\nu}_\mu + 2\nu_\mu$

  (c) Charge: $-1 \to +1 - 3 + 1 = -1$ conserved

  Baryon number: $1 \to 1 + 0 + 0 + 0 + 0 + 0 + 0 = 1$ conserved

  Lepton number:

        (i) electrons: $0 \to 0 + 3 - 1 - 3 + 1 + 0 + 0 = 0$ conserved

        (ii) muons: $0 \to 0 + 0 + 0 + 0 + 0 - 2 + 2 = 0$ conserved

(Problem 13-45 continued)

Strangeness: $-3 \to 0+0+0+0+0+0+0 = 0$ not conserved

Overall reaction has $\Delta S = +3$; however, none of the individual reactions exceeds $\Delta S = +1$, so they can proceed via the weak interaction.

13-49. (a) $\Delta t = t_2 - t_1 = \dfrac{x}{u_2} - \dfrac{x}{u_1} = \dfrac{x(u_1 - u_2)}{u_1 u_2}$ Note that $u_1 u_2 \approx c^2$

$$\Delta t \approx \dfrac{x(u_1 - u_2)}{c^2} = \dfrac{x\,\Delta u}{c^2}$$

(b) $E = \dfrac{mc^2}{\sqrt{1 - u^2/c^2}}$ (Equation 2-10). Thus, $\dfrac{u}{c} = \left[\dfrac{1 - (m_o c^2)^2}{E^2}\right]^{1/2} \approx 1 - \dfrac{1}{2}\left(\dfrac{m_o c^2}{E}\right)^2$

(c) $u_1 - u_2 = c\left[1 - \dfrac{1}{2}\left(\dfrac{m_o c^2}{E_1}\right)^2 - 1 + \dfrac{1}{2}\left(\dfrac{m_o c^2}{E}\right)^2\right]$

$$= \dfrac{c}{2}\left(\dfrac{m_o c^2}{E_2}\right)^2 - \dfrac{c}{2}\left(\dfrac{m_o c^2}{E_1}\right)^2 = \dfrac{c(m_o c^2)^2}{2}\left[\dfrac{E_1^2 - E_2^2}{E_1^2 E_2^2}\right]$$

$$= \dfrac{c(20\,eV)^2}{2}\left[\dfrac{(20\times 10^6\,eV)^2 - (5\times 10^6\,eV)^2}{(20\times 10^6\,eV)^2(5\times 10^6\,eV)^2}\right]$$

$$= \dfrac{c(20\,eV)^2}{2}\left[\dfrac{(20)^2 - (5)^2}{(20)^2(5)^2(10^6\,eV)^2}\right] = 7.5\times 10^{-12}\,c$$

$$\Delta T \approx \dfrac{x\,\Delta u}{c^2} = \dfrac{(170{,}000\,c\cdot y)(7.50\times 10^{-12}\,C)}{c^2} = 1.28\times 10^{-6}\,y = 40.3\,s$$

(d) If the neutrino rest energy is 40 eV, then $\Delta u = 3.00\times 10^{-11}\,c$ and $\Delta t \approx 161\,s$. The difference in arrival times can thus be used to set an upper limit on the neutrino's mass.

Chapter 14 – Astrophysics and Cosmology

14-1.

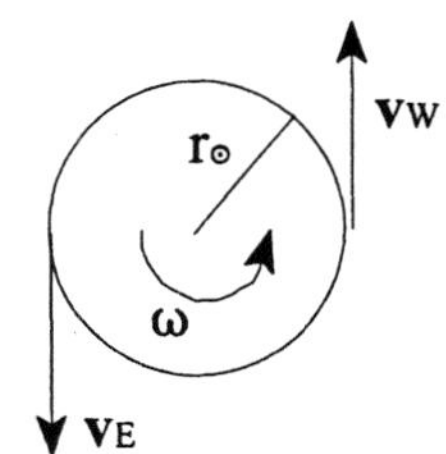

$|v_W - v_E| = 4$ km/s. Assuming Sun's rotation to be uniform, so that $v_W = -v_E$, then $|v_W| = |v_E| = 2$ km/s.

Because $v = 2\pi r/T$, $v_E = 2\pi r_\odot / T$ or

$$T = \frac{2\pi r_\odot}{v_E} = \frac{2\pi(6.96\times 10^5\ km)}{2\ km/s} = 2.19\times 10^6\ s = 25.3\ days$$

14-5. Observed mass (average) $\approx$ 1 H atom/m^3 = 1.67×10^{-27} kg/m^3 = 10% of total mass

$\therefore$ missing mass = $9\times 1.67\times 10^{-27}\ kg/m^3 = 1.50\times 10^{-26}\ kg/m^3$

500 v/cm^3 = 500 $\times$ 10^6 v/m^3,

so the mass of each v would be $= \dfrac{1.50\times 10^{-26}\ kg/m^3}{500\times 10^6\ v/m^3} = 3.01\times 10^{-35}\ kg$

or $m_v = \dfrac{3.01\times 10^{-35}\ kg}{1.60\times 10^{-19}\ J/eV} \times c^2\left(\dfrac{m^2}{s^2}\right) = 16.9\ eV$

14-9. (a) $M = 0.3\ M_\odot$ $T_e = 3300\ K$ $L = 5\times 10^{-2} L_\odot = 1.93\times 10^{25}\ W$

    (b) $M = 3.0\ M_\odot$ $T_e = 13,500\ K$ $L = 10^2 L_\odot = 3.85\times 10^{28}\ W$

    (c) $R \sim M \;\Rightarrow\; R = \alpha M \;\Rightarrow\; \alpha = R_\odot / M_\odot$

$$R_{0.3} = \alpha(0.3\,M_\odot) = \frac{R_\odot}{M_\odot}(0.3\,M_\odot) = 0.3\,R_\odot = 2.09\times 10^8\ m \quad \text{Similarly, } R_{3.0} = 3.0\,R_\odot = 2.09\times 10^9\ m$$

$$t_L \sim M^{-3} \;\Rightarrow\; t_L = \beta M^{-3} \;\Rightarrow\; \beta = t_{L\odot}/M_\odot^{-3} = t_{L\odot} M_\odot^3$$

    (d)

$$t_L(0.3) = \beta\,(0.3\,M_\odot)^{-3} = t_{L\odot} M_\odot^3\,(0.3\,M_\odot)^{-3} = (0.3)^{-3}\,t_{L\odot}$$

$$\text{or} \quad t_L(0.3) = 37\,t_{L\odot}. \quad \text{Similarly, } t_L(3.0) = 0.04\,t_{L\odot}$$

14-13. $R_S = 2GM/c^2$ (Equation 14-24)

(a) Sun $R_S = 2 \times 6.67 \times 10^{-11} \times 1.99 \times 10^{30}/c^2 = 2.9 \times 10^3 \, m \approx 3 \, km$

(b) Jupiter $(m_J = 318 \, m_E)$ $R_S = 2.8 \, m$

(c) Earth $R_S = 8.86 \times 10^{-3} \, m$ $(\approx 9 \text{ mm!})$

14-17. The process that generated the increase could propagate across the core at a maximum rate of c, thus the core can be at most

$1.5 \, y \times 3.15 \times 10^7 \, s/y \times 3.0 \times 10^8 \, m/s = 1.42 \times 10^{16} \, m = 9.5 \times 10^4 \, AU$ in diameter. The Milky Way diameter is $\approx 60{,}000 \, c{\cdot}y = 3.8 \times 10^9 \, AU.$

14-21. Wien's law (Equation 3-11): $\lambda_{max} = \dfrac{2.898 \, mm \cdot K}{T} = \dfrac{2.898 \, mm \cdot K}{2.728 \, K} = 1.062 \, mm$

(this is in the microwave region of the EM spectrum)

14-25. SN1987A is in the Large Magellanic cloud, which is 170,000 c·y away; therefore (a) supernova occurred 170,000 years BP.

(b) $E = K + m_o c^2 = \dfrac{m_o c^2}{\sqrt{1 - \dfrac{v^2}{c^2}}}$

$$m_o^2 = 9.38 \times 10^8 \, eV \;\;\Rightarrow\;\; 10^9 + 9.38 \times 10^8$$

$K = 10^9 \, eV,$ $\qquad = \dfrac{9.38 \times 10^8}{\sqrt{1 - \dfrac{v^2}{c^2}}}$ or $v = 0.875 \, c$

Therefore, the distance protons have traveled in $170{,}000 \, y = v \times 170{,}000 \, y = 149{,}000 \, c{\cdot}y.$ No, they are not here yet.

14-29.

$$E = \frac{1}{2}mv^2 + \left(-GmM_\odot/r\right) \qquad F_G = GM_\odot m/r^2 = mv^2/r$$

$$\text{or} \quad GM_\odot m/r = mv^2 \quad \Rightarrow \quad \frac{1}{2}mv^2 = \frac{1}{2}GM_\odot m/r$$

$$\therefore \quad E = \frac{1}{2}\frac{GM_\odot m}{r} + \left(-\frac{GM_\odot m}{r}\right) = \frac{1}{2}\left(-\frac{GM_\odot m}{r}\right)$$

14-33. Earth is currently in thermal equilibrium with surface temperature ≈ 300 K. Assuming Earth radiates as a blackbody $I = \sigma T^4$ and $I = \sigma(300)^4 = 459\ W/m^2$. The solar constant $f = 1.36 \times 10^3\ W/m^2$ currently, so Earth absorbs $459/1360 = 0.338$ of incident solar energy. When $L_\odot \;\rightarrow\; 10^2 L_\odot$ then $f = 10^2 f$. If the Earth remains in equilibrium.

$I = 0.338 \times 1.36 \times 10^3 \times 10^2 = \sigma T^4$ or $T = 994\ K = 676\ °C$ sufficient to boil the oceans away.

However, the v_{rms} for H_2O molecules at 994 K is $v_{rms} = \sqrt{3RT/M} = \sqrt{\dfrac{3 \times 8.31 \times 949}{18 \times 10^{-3}}} =$

1146 m/s = 1.15 km/s. The v_{esc} = 11.2 km/s (see solution to problem 14-26). Because $v_{rms} \approx 0.1\ v_{esc}$, the H_2O will remain in the atmosphere.